D1282777

THE COSMIC WEB

THE COSMIC WEB

Scientific Field Models and Literary Strategies in the Twentieth Century

N. KATHERINE HAYLES

CORNELL UNIVERSITY PRESS

ITHACA AND LONDON

First published 1984 by Cornell University Press.
Published in the United Kingdom by
Cornell University Press Ltd., London.

International Standard Book Number 0-8014-1742-2
Library of Congress Catalog Card Number 84-45141
Printed in the United States of America
Librarians: Library of Congress cataloging information appears on the last page of the book.

The paper in this book is acid-free and meets the guidelines for permanence and durability of the Committee on Production Guidelines for Book Longevity of the Council on Library Resources.

For Trurl

CONTENTS

PREFACE

WE ARE LIVING AMID the most important conceptual revolution since Copernicus argued that the earth was not the center of the universe. Of this new revolution much has been written; but most of the discourse about it has taken place within the boundaries of a single discipline, so that there exists no accepted appellation for it and no clear definition of it as a general cultural phenomenon. In this book I have singled out the "field concept" as the theme that is at the heart of this revolution, and have examined its various manifestations in the models of physics and mathematics, the theories of the philosophy of science and linguistics, and the structure and strategies of literary texts.

The *field concept,* as I use the term, is not identical with any single field formulation in science. For the men and women who work with the various scientific field models and theories from day to day, they have highly specific meanings and applications. The term "field concept," by contrast, draws from many different models those features that are isomorphic, and hence that are characteristic of twentieth-century thought in general. The only way to approach a satisfactory understanding of the field concept is to examine and compare a wide range of phenomena that embody it; and that is the major burden of this book. But the more salient features of the field concept can be sketched here.

Perhaps most essential to the field concept is the notion that things are *interconnected.* The most rigorous formulations of this idea are found in modern physics. In marked contrast to the atomistic Newtonian idea of reality, in which physical objects are discrete and events are capable of occurring independently of one another and the observer, a field view of reality pictures objects, events, and observer as belonging inextricably to the same field; the disposition of each, in this view, is influenced—sometimes dramatically, sometimes subtly, but in every instance—by the disposition of the others.

Another aspect of the field concept, one that figures importantly in several disciplines and in the works of the five authors studied in this book, is the notion of the *self-referentiality* of language. Because everything, in the field view, is connected to everything else by means of the mediating field, the autonomy assigned to individual events by language is illusory. When the field is seen to be inseparable from language, the situation becomes even more complex, for then every statement potentially refers to every other statement, including itself. This implication of the field concept is central to the literary responses to the field view that I explore in this book.

The field concept cannot, of course, be summed up so briefly. Like many other general terms—"democracy," "romanticism," "modernity"—it can be fully understood only by association with the various phenomena it is used to describe. Studying some of the embodiments of the field concept is my purpose here. The metaphor I have chosen to represent the world as construed by the field concept is the "cosmic web." That metaphor communicates something of the interconnectedness and "stickiness" of self-reference of which I have already spoken. Its other applications are discussed in the Introduction and illustrated in the following chapters.

One of the many ideas that the field view revises is the notion of a one-way chain of reaction between the event labeled as the "cause" and that labeled the "effect." Although I spoke earlier of the "influence" of the field view on modern literature, I do not mean to imply by this that the literature I discuss is "caused" by scientific field models. Rather, the literature is an imaginative response to complexities and ambiguities that are implicit in the models but that are often not explicitly recognized. Thus a comprehensive picture of the field concept is more likely to emerge from the literature and from science viewed together than from either one alone. In this sense the literature is as much an influence on the scientific models as the models are on the literature, for both affect our understanding of what the field concept means in its totality.

The Introduction and Chapter 1 lay out the conceptual framework for the book. The literary chapters that follow have been arranged in ascending order of the complexity of the authors' resistances to the field concept; none of these writers adopts the field concept simplemindedly or wholeheartedly. The chapter titles indicate the nature of these strategies: Pirsig propagandizes the field concept as a way to heightened consciousness, Lawrence and Nabokov are ambivalent toward it, Borges subjects it to irreverent subversive transformations, and Pynchon sees it as an unavoidable, and ultimately tragic, double-bind.

Readers will be likely to find omitted from this list of authors favorites that they think should have been included. Indeed, if my argument is sound, there must be many texts manifesting the influence of the field concept, in many different disciplines. I have thought it best to treat a few models and authors in depth rather than mention many superficially. Readers who find these treatments interesting or persuasive can test the argument further with texts of their choice. My purpose is to blaze a trail rather than cover the terrain. If this book serves to open a dialogue, I am content; I leave it to others to complete the exchange.

Many friends and colleagues have given generously of their time to read the manuscript and offer suggestions, and I am deeply appreciative of their help. They include W. T. Jones, David Smith, Noel Perrin, Thomas Vargish, Delos Mook III, Christine Salada, Nancy Frankenberry, and Nancy Crumbine. I am also grateful to the students in my Science and Literature seminars who gave me encouragement and helped keep me going, and to colleagues in the Science and Humanities seminar at the California Institute of Technology who gave me a hard time and helped keep me honest. Anonymous readers for Cornell University Press provided many helpful suggestions, from which I benefited greatly.

I am also grateful to Dartmouth College for a Faculty Fellowship in the fall of 1979, during which a start was made on this book, and to the National Endowment for the Humanities for a fellowship during 1979–1980, which allowed me to complete the original draft. The Humanities and Social Sciences Division of the California Institute of Technology provided secretarial support and an office during the tenure of my NEH fellowship. I am grateful to the University of Missouri–Rolla for a Weldon Spring Faculty Grant in the summer of 1983 that enabled me to work on revisions.

Chapters 3, 4, and 6 of this book appeared in slightly different form as articles in *Mosaic: A Journal for the Interdisciplinary Study of Literature* (September 1982), *Contemporary Literature* (Winter 1982), and *The Markham Review* (Summer 1983), respectively. I am grateful to the publishers and editors for permission to use this material. For permission to quote from Robert M. Pirsig, *Zen and the Art of Motorcycle Maintenance* (copyright © 1974 by Robert M. Pirsig), I am grateful to the publishers, William Morrow & Company, Inc., New York, and The Bodley Head Ltd., London.

Final thanks goes to Terry Viens, who read the manuscript, offered suggestions and criticisms, and helped prepare it for publication. Without her invaluable help, advice, and support, I could not have completed this project.

N. Katherine Hayles

Rolla, Missouri

THE COSMIC WEB

Tout se tient.

Ferdinand de Saussure

What we observe is not nature in itself but nature exposed to our method of questioning.

Werner Heisenberg

. . . the grammatical background of our mother tongue . . . includes not only our way of constructing propositions but the way we dissect nature and break up the flux of experience into objects and entities to construct propositions about.

Benjamin Whorf

We are suspended in language.

Niels Bohr

Everything is connected.

Thomas Pynchon, *Gravity's Rainbow*

INTRODUCTION

THE TWENTIETH CENTURY has seen a profound transformation in the ground of its thought, a change catalyzed and validated by relativity theory, quantum mechanics, and particle physics. But the shift in perspective is by no means confined to physics; analogous developments have occurred in a number of disciplines, among them philosophy, linguistics, mathematics, and literature. From the vantage of the closing decades of this century, the appearance of a Copernican revolution sweeping through the culture is irresistible. I shall speak of it as a revolution in world view. The people most responsible for the transformation did not necessarily consider themselves part of a larger movement; nevertheless, their streams of inquiry flowed in a similar direction, the converging courses of which changed the intellectual terrain of modern thought.

The essence of the change is implicit in the heuristic models adopted to explain it. Characteristic metaphors are a "cosmic dance," a "network of events," and an "energy field." A dance, a network, a field—the phrases imply a reality that has no detachable parts, indeed no enduring, unchanging parts at all. Composed not of particles but of "events," it is in constant motion, rendered dynamic by interactions that are simultaneously affecting each other. As the "dance" metaphor implies, its harmonious, rhythmic patterns of motion include the observer as an integral participant. Its distinguishing characteristics, then, are its fluid, dynamic nature, the inclusion of the observer, the absence of detachable parts, and the mutuality of component interactions.

This concept is very different from the older paradigm implicit in

Newtonian mechanics, the atomistic, "common sense" perspective we are all familiar with that views the world as composed of objects situated in an empty, rectilinear space and moving through time in one direction. The intuitive obviousness of this view to us is no doubt reinforced, as Benjamin Whorf has suggested, by the deep structure of Indo-European languages, which embodies its fundamental assumptions: the separation between subject and object, the duration of objects through time, and the uniform, unidirectional flow of time.[1] But we should not lose sight of the fact that the scientific expression of this view is a relatively recent phenomenon, dating from the latter seventeenth century. Since its beginning as a scientific world view can be historically determined, its ending perhaps can too. Although it is still the view most of us hold, there are indications that its decline has already begun.

The quantum field theories of high energy physics, for example, can lead to a very different perspective. Some physicists, faced with the dazzlingly rapid transformations that subatomic particles undergo, have suggested that it is more economical to think of the essential entity not as the particle, but as the underlying quantized field. In this view "particles" are expressions of the field's conformation at a given instant, appearing as the field becomes concentrated at one point and disappearing as it thins out at another. Particles are not to be regarded as discrete entities, then, but rather (in Hermann Weyl's phrase) as "energy knots."[2] What the particle was for the Newtonian paradigm, the field is for the new paradigm.

Humanistic disciplines also reflect this change in view, as can be seen by comparing the dominant metaphors of our era with those associated with previous paradigms. When the eighteenth-century rationalists imaged the world as a clock, for example, they implied that the world was composed of interlocking parts, that the parts could be detached from one another, and that an intelligent observer could deduce the function

[1]Benjamin Whorf, *Collected Papers on Metalinguistics* (Washington, D.C.: Foreign Service Institute, 1952), pp. 3–8, 27–52. The problem with Whorf's thesis is that in some strains of European culture (for example, alchemical thought in sixteenth- and seventeenth-century England), very different world views have emerged, even though the language was essentially the same as that Newton spoke. Clearly other cultural factors, in addition to the deep structure of language, are responsible for the dominance of the Newtonian world view from the seventeenth through the nineteenth centuries.

[2]Hermann Weyl, *Philosophy of Mathematics and Natural Science* (Princeton: Princeton University Press, 1949), p. 171.

of the machine from the workings of its parts. These assumptions were more or less conscious and intentional. The clock metaphor, however, also implied other attributes not foregrounded in consciousness, but nevertheless capable of affecting (or expressing) unconscious expectations about the nature of reality. Among these implications was the inference that the world, like a machine, had a fixed and static form and once set in motion would run itself without further need of divine intervention. Closer to the surface was the premise that its workings were rational, and that the proper way to investigate it was through the linear chains of inductive and deductive reasoning for which Bacon had argued in the *Novum Organon*.

In contrast to these eighteenth-century expectations were those implicit in the nineteenth-century Romantic image of the world as an organism. In his discussion of Romanticism, Hans Eichner points out that the difference between the two metaphors ("Machines do not grow, organisms do") was reflected in many different areas of the culture.[3] In general, the eighteenth-century emphasis on static categories changed in the nineteenth century to interest in the dynamics of change. Taxonomy yielded to evolution in biology, poetics to history in literary theory, and mimesis to a literature of interiority. Moreover, as Eichner observes, if the world is a dynamic, living whole, it cannot be entirely understood through reason alone. The mysterious essence of life requires for its understanding the sympathetic imagination. It is possible to take a machine apart and examine it without imperiling its function; indeed, its end can be most clearly understood if it is divided into parts. But if a living being is dissected, the essential quality of life is destroyed; the remaining parts will never add up to the original whole. There was thus a sense among the Romantics that the whole is something other than the sum of its parts, and this "otherness" was identified with the life force.

The twentieth-century metaphor of the "cosmic dance" has in common with the Romantic metaphor of an "organism" the implication that the whole cannot be adequately represented as the sum of its parts, and the emphasis on the dynamic, fluid nature of reality. But whereas the Romantics identified this dynamism with a specifically *living* force,

[3]Hans Eichner, "The Rise of Modern Science and the Genesis of Romanticism," *PMLA*, 97 (1982), 8–30.

the modern period links it with a breakdown of universal objectivity. This is the difference, for example, between Henri Bergson's theory of duration, which grows out of his notion of the *élan vital,* and Einstein's concept of time in the Special Theory of Relativity. Bergson's time is flexible because it is associated with a life force that perceives in a nonmechanical way; for Einstein, time is relativistic not because the universe is infused with life, but because the motion of the observer affects the language of description. In Einstein's theory, this qualification does not depend on the observer being animate; the same result would be obtained if the measurement were made by a nuclear decay clock rather than a person.

So far I have been emphasizing the differences between the contemporary world model and Romanticism. Are there no continuities? Is Romanticism a sport in the history of modern thought, a deviation from the otherwise steadily increasing rationalism of a scientific age? This view, which Eichner espouses, depends upon the premise that science in the twentieth century is not essentially different in its philosophical assumptions from science in the 1700s. "If Galileo could be hijacked by a time machine, taught English, and dropped into contemporary Boston," Eichner asserts, "he would . . . feel completely at home at M.I.T. Schelling would have to be brainwashed."[4] But Galileo would require far more than a crash course in mathematics to become acclimated to twentieth-century science. He would also have to abandon a belief in strict causality and accept the idea that our "particles," rather than existing as collections of enduring, definitive objects, manifest themselves as "tendencies to exist." Perhaps most disturbing to the time-transported Galileo would be the notion that a strict separation between subject and object is not possible and that, accordingly, there are inherent limits on how complete our knowledge of any physical system can be. These twentieth-century epistemological assumptions have more in common with Romanticism than they do with seventeenth-century science. Modern science is not renouncing Romanticism, only changing its emphases. If we were to try to graph the relationship between these eras it would not be, as Eichner proposes, a straight line from eighteenth-century rationalism to twentieth-century positivism with Romanticism as a deviant point, but a curve that, by

[4]Ibid., p. 24.

including Romanticism, thereby proceeds in a radically altered direction.[5]

One of the important points of continuity between Romanticism and the field concept is the appearance of inherent limits on sequential, logical analysis. In physics, the limit emerges as an upper bound on what can be expressed about reality. Why these limitations occur will be more fully and rigorously explored in the next chapter, when some of the important scientific models that lead to this conclusion will be examined. For the moment, they can be understood intuitively by considering what the nature of the universe would be if it were participating in a cosmic dance.

Imagine, for example, that we are sitting in a diner, waiting for a hamburger. In the ordinary view the plate, knife, fork, and ketchup bottle are "real," while the pattern they form is a transitory artifact of their relative positions. But suppose that we were to shift our perspective so that we regarded the *pattern* as "real," and the ketchup bottle, plate, knife, and fork as merely temporary manifestations of that particular pattern. This radically altered perspective is analogous to the shift in view suggested by quantum field theory, and is what Fritjof Capra, a particle physicist and prophet of the holistic world view, has in mind when he asserts that "the whole universe appears as a dynamic web of inseparable energy patterns."[6]

The resistance of our language to expressing this view can scarcely be overestimated. If we try to construct an objective description of it, the difficulties will quickly become apparent. First of all, such a description must proceed from a point within the "dynamic web," for if the dance *is* the universe, there is no point outside it. Imagine, then, attempting an internal, causal description of these "events." As one configuration shifts to another and as "particles" appear or disappear in response to the field as a whole, the usual distinction between cause and effect breaks down because linear sequences of causality depend upon being able to define a one-way interaction between the event regarded as a "cause" and that considered as an "effect." But when the interaction is

[5]Eichner acknowledges that Galileo would have a few uneasy moments wrestling with Heisenberg's Uncertainty Principle. In my opinion, this formulation trivializes the epistemological issues that are involved. The extent and depth of these issues will be explored in the next chapter.

[6]Fritjof Capra, *The Tao of Physics* (New York: Bantam Books, 1977), p. 69.

multidirectional—when every cause is simultaneously an effect, and every effect is also a cause—the language of cause and effect is inadequate to convey the mutuality of the interaction. Causal descriptions will not do because causal terminology implies a one-way interaction that falsifies the essence of what we want to convey.

Suppose instead we try a metaphor, or perhaps I should say, a different metaphor: a constantly turning kaleidoscope whose shifting patterns arise from the continuing, mutual interaction of all its parts. Two restrictions to a complete description then become apparent. Because we cannot describe the totality of the dance, which is incessant and infinite, we must stop the kaleidoscope in our imaginations, calling each slice-of-time configuration a "pattern." But by stopping the kaleidoscope we have lost the dynamic essence of the dance, for the static "patterns" never in fact existed as discrete entities. The problem is endemic to synchronic analyses; any finite, slice-of-time model will encounter the same problem. One set of limitations thus emerges from the dynamic nature of this reality.

A second group of limitations derives from the lack of an exterior, "objective" point from which to observe. No matter where we stand we are within the kaleidoscope, turning with it, so that what we see depends on where we stand. To change positions does not solve the problem, because the patterns are constantly changing: what we see when we change positions is not what we would have seen, for in the intervening time the patterns will have changed, and our shift in position will be part of that change. Moreover, there will always be one place we can never see at all—the spot we are standing on. Like the figure in a painting who wishes to gesture toward the picture that contains him, we can never arrive at a complete and unambiguous description of this reality because we are involved in what we would describe. To posit such a reality is inevitably to encounter these limitations, because its essence, its all-of-a-piece dynamic wholeness, is what causes the limitations to occur.

So far I have been speaking of the obstacles to a complete description that occur when we try to use a natural language. But the difficulties are more general than this; as we shall see in the next chapter, they also appear when one attempts a complete description in scientific and mathematical languages. The realization that there are inherent limits on what can be spoken, and that these limits arise because language is part of the field being described, is at the heart of the revolution implicit in a field concept of reality. The stickiness of this situation, our inability to extricate the object of our description from the description itself, suggests that a more appropriate image for the field concept than the "cosmic dance" is the "cosmic web."

A central metaphor in this study, the cosmic web has connotations worth exploring. Readers who desire a visual image to go along with the metaphor may imagine it as a network of strands coextensive with space. Note that the web is not space itself, nor does it "contain" space. Rather it is an artifact, a created object whose artificiality corresponds to the conceptualization of the field models it signifies; what we are concerned with in these models is not reality as such, but conceptualizations that may or may not correspond with whatever we call reality. Imagine further that the web is composed of articulated joints, much as a spider's web is. These joinings will serve as a convenient reminder that the verbal models we shall be examining are also articulated, in the double sense of being utterances and of being composed of discrete units joined together. Once the web is constructed, these joinings may stand for, or gesture toward, a seamless whole; but this evocation can be attempted only through a medium that is itself linear, sequential, and articulated. The prey the cosmic web is designed to entrap is the dynamic, holistic reality implied by the field concept. But the prey always escapes, precisely because the web is articulated; as we shall see, to speak is to create, or presuppose, the separation between subject and object that the reality would deny. What is captured by the cosmic web is thus not the elusive whole, but the observer who would speak that whole. Hence the cosmic web is inherently paradoxical, deriving its deepest meaning from a whole that it can neither contain nor express. Its history can be told as the history of certain paradoxes.

To enumerate these paradoxes is to begin to realize the scope of the paradigm shift which has brought them into focus. Since any statement in a field model can be made to refer to itself if the statement is part of the field that the model posits, statements have the potential to become self-referential, a realization as central to Gödel's theorem as it is to Borges's fictions. The supposition that there is a speaking subject separate from the object that is being spoken about also becomes problematic, and generates an uneasiness that is as apparent in most modern interpretations of the Uncertainty Relation as in Pynchon's *Gravity's*

Rainbow. Another assumption that becomes paradoxical in a field model is the premise that it is possible to establish an unambiguous time-line for spatially separated events, a conception whose unraveling is as important to relativity theory as it is to Nabokov's *Ada.*

These brief references are meant to give the reader a sense of how developments in a number of different disciplines can be related to the emergence of modern field models; they will be discussed in more detail in the following chapters. This is primarily a study about literature, however, and my major emphasis is on how literary theory and form have been shaped by the change of paradigms. The groundwork for a field view of language was laid in 1916 with the posthumous publication of Ferdinand de Saussure's *Cours de linguistique générale.* In proposing *la langue* as a proper subject for linguistics, Saussure argued that language systems should not be regarded as collections of discrete semantic units, but as unified systems in which meaning derives from the relational exchanges between signs. The effect of this view was to locate meaning not in a one-to-one correlation between the sign and its external referent, but in the relations between signs. When Saussure argued that the entire linguistic structure changes with the addition or omission of a single lexical unit, he conceived of language as an integrated, nondivisible whole, that is to say, as a unified field composed of parts but not reducible to the sum of its parts.

That Saussure's proposals are remarkably similar in spirit to those occurring about the same time in physics and mathematics does not require that Saussure knew of Einstein's 1905 papers or read *Principia Mathematica.* Indeed, to suppose that such parallels require direct lines of influence is to be wedded to the very notions of causality that a field model renders obsolete. A more accurate and appropriate model for such parallel developments would be a field notion of culture, a societal matrix which consists (in Whitehead's phrase) of a "climate of opinion" that makes some questions interesting to pursue and renders others uninteresting or irrelevant. Such a field theory of culture has yet to be definitively articulated, and is beyond the scope of this study. But it is already possible to see some of the elements it would include. It would, for example, define more fully how a "climate of opinion" is established, and demonstrate that it is this climate, rather than direct borrowing or transmission, that is the underlying force guiding intellectual inquiry. This climate would be, of course, as capable of influencing scientific inquiry as it is of guiding any other conceptualization. Such a history would insist that we not be misled by a causal perspective into thinking of correspondences between disciplines as one-way exchanges, for example, by asserting that the change in scientific paradigms *caused* a shift in literary form. In a field model, the interactions are always mutual: the cultural matrix guides individual inquiry at the same time that the inquiry helps to form, or transform, the matrix.

In its treatment of the modern novel, this history would show that the cultural matrix was so configured as to draw modern novelists to considerations similar to those Saussure entertained. It would point out, for example, that just as linguistic meaning in a field model was deemed to derive from relational exchanges within the language system, so meaning in a literary text was deemed to derive not from a mimetic relationship between the text and "real life," but from the internal relations of literary codes. It could then show that explorations of this possibility in the novel proceeded in two different but related directions. One turned inward, assuming that literature, like language, is an internal system that has no necessary reference to anything outside itself. In extreme form, this train of thought resulted in a literature that is both nonreferential and solipsistic. One thinks, for example, of the narrator of Beckett's *The Unnamable,* whose connections with external reality have been progressively stripped away until there is finally not even a truncated body attached to the voice; all that exists is the voice, speaking to itself. Because this inward-turning literature is nonmimetic in its orientation, the term "anti-realism" can properly be applied to it.

Our supposed history could then go on to show that the "anti-realism" rubric often includes other narratives whose orientation is, however, quite different. These texts, although they may possess "anti-realistic" traits, turn outward toward an apparently external referent. The nature of the reality being represented is, however, radically altered, for it is no longer simply external and objective, nor is it represented as an object separate and distinct from its verbal expression. Rather, it is assumed to be continuous with the text, interpenetrating the signifiers that re-present it. A conservative example is Conrad's *The Heart of Darkness,* in which external reality is filtered first through the narrator, then through the internal perceptions of the protagonists, so that the meaning exists, as the narrator asserts of Marlowe's storytelling, not as the kernel of a nut but as a kind of luminous haze without a definitive locus in the signifiers themselves. Here two implications of the field concept come into play: that the whole is composed of parts but cannot be reduced to them; and that the observers are an inextricable part of the field. In more radical versions the external reality, though putatively existing, is irrecoverable, for the subject's perceptions of it have so deformed and merged with it as to eradicate the possibility of recovery; one thinks here, for example, of Faulkner's *The Sound and the Fury*.

Our history could further demonstrate that the impulse to represent a continuous reality need not necessarily be expressed as extreme subjectivity. Also possible are literary texts that try to re-create the continuum within the text. This immediately involves the author in paradoxes of self-referentiality, for the enabling premise that the text is part of the whole also implies that the whole can be contained within the part, leading to the infinite regress of a part containing a whole within which is contained the part. . . . Familiar examples here include many of Borges's fictions.

My purpose in sketching these possibilities obviously is not to write this history, but to show that it is possible, and to suggest some possible points of reference. The present study begins with the premise that such a history would end by establishing: that well-known developments in the modern novel are part of a larger paradigm shift within the culture to the field concept. Rather than attempt this history, I have assumed it by locating a group of representative novels within a larger cultural context that includes physics, mathematics, and philosophy. By demonstrating the usefulness of the premise in understanding these texts, I hope to encourage further work that would undertake to explore the premise.

Since I am assuming that these novels are affected by the shift toward the new paradigm, the reader may wonder whether I also mean to imply that the authors are thoroughly conversant with field models, or are trying to re-create it within their works. In my opinion, both of these models of "influence" oversimplify the interaction between an author and his culture. Most of the authors I am concerned with know little of science, and what little they do know is often colored by their idiosyncratic interpretations. In addition, most of them write for a small literary audience, and this further helps to insulate them from developments in science. With few exceptions, these authors are reacting not to science as such, but to a more general set of ideas pervasive in the culture. One purpose of this book is to provide readers from both sides of the cultural divide with the information they need to see that the connecting link between these ideas is the field concept, and to demonstrate that it is as capable of informing literary strategies as it is of forming scientific models.

Given the hundreds of literary texts that might be studied, what has governed my choice? There can be no question of choosing the "right" texts, for the argument is that the influence of the field concept is pervasive throughout the literature. My selection was guided by two criteria: first, I wanted texts that would reveal how wide the range is of literary strategies that can emerge from an author's encounter with the field concept; and second, I wanted texts that would evidence varying degrees of knowledge and sympathy toward science. Lawrence and Nabokov know little about the science, whereas Pynchon knows a great deal; Lawrence mostly dislikes what little he knows, while Borges delights in modern set theory and reads mathematical texts to learn more. The selection is diverse enough to show that a writer does not have to be post-modernist to be affected by the field concept; authors as different in their literary techniques and philosophies as Lawrence and Nabokov, Pynchon and Pirsig, are all affected.

The desire to show the full complexity and range of response has also dictated the book's organization, for the literary chapters are arranged according to the authors' increasing resistance to the field concept. Robert Pirsig's *Zen and the Art of Motorcycle Maintenance* comes first because he seeks most wholeheartedly to embrace it. It is interesting that few of Pirsig's sources come directly from physics; Einstein is merely mentioned, for example, and Heisenberg appears not at all. Consequently, this text also demonstrates that a writer can be concerned with issues that have been brought into focus by the paradigm shift without necessarily being familiar with those sources that most directly brought it about. Pirsig's book is the one work of this study that has not been incorporated into the literary canon, finding its audience in the mass market rather than among a literary coterie. It is thus an important text for demonstrating how a set of ideas can be broadcast through the culture, transforming it in turn. From this popular treatment of the field concept emerges a question that haunts all of the writers in this study: can the representation of a holistic field be accomplished within the linear flow of words, or is the attempt inherently limited by the fragmentation of the medium?

Lawrence and Nabokov come next because they demonstrate how writers who are relatively ignorant of the new science nevertheless participate in the cultural matrix and so, willy-nilly, encounter in some form the matrix's underlying paradigm. Nabokov and Lawrence are further tied together by their ambivalence toward the new science; neither is fond of the theoretical sciences, yet both find in it propositions that they wish to appropriate for their own ends. Lawrence in particular had only the foggiest notion of what relativity was, but he knew enough to sense that the old ways of looking at the world were crumbling, and into the gap he meant to insert his own version of a field model. Working partly from ignorance, partly from intuition, and partly from Bergsonian theory, Lawrence proposed a "subjective science," at the center of which is a psycho-physical model that unites subject and object into one pulsating, dynamic field. But the only way Lawrence could envision this unified field was as two polarities locked together into tense opposition in highly unstable configurations. As a result, this holistic "field," which Lawrence identifies with the unconscious, keeps fragmenting and reforming, only to break apart again. Ironically, the attempt to sustain the field fails not because Lawrence finds it impossible to represent, but because he fears what such a reality would entail if it were represented. Lawrence's fiction thus evolves in dialectical fashion from two world views, the old in his opinion moribund, but the new too fraught with danger for him to sustain it.

Unlike Lawrence, Nabokov chooses to ignore the psychological implications of a field model, concentrating instead on the one implication of this model that he finds attractive: that time may be reversible. Informed by Nabokov's ambivalence toward the field model, *Ada* has inscribed within it two contradictory impulses: the desire to move into the future, whence comes the scientific validation for reversible time; and the nostalgic wish to recover the past—an enterprise which, if the theory is correct, should be possible. The problem is how to represent both of these impulses at once, since they point in opposite directions. Nabokov's solution is to imagine twinned worlds, Terra and Antiterra, with a not-quite-perfect alignment between them that keeps them from canceling each other out. The dialectic of Nabokov's fiction is thus between the deterministic past of the Newtonian world view and the reversible future of post-Einsteinian relativity, between the confinement of a static, predictable space and the free-wheeling permutations of a synchronous field.

Compared to the impressionistic way in which Lawrence and Nabokov interpret a field model, Borges's response is extremely precise, though no less problematic. What fascinates Borges is the prospect of a set that contains itself, a whole that both contains and is contained by the part. Such paradoxes are implicit in many representations of field models, because the representation is at once the whole, in the sense that it images the field, and the part, in the sense that it is contained within the whole it figures. This paradox, central to Borges's fictions, is explored through the infinite sets and transfinite numbers of Cantor set theory. Borges's assumption is that the Newtonian universe must crumble when confronted with the antinomies to which this theory gave rise. But he does not want a new reality to come into being either. Rather, he juxtaposes the new "loss of certainty" with old certainties to render everything uncertain. In this chapter the new world model engages the old not so much in a dialectic as in a collision that subverts both. Borges's response to the field concept is thus essentially a strategy of subversion.

At the center of Pynchon's *Gravity's Rainbow* is the question with which we began—can a holistic field be represented in a linear flow of words?—and his treatment of it is formidably complex. His exploration of its implications includes meditations on the indeterminacy of the new physics, speculations on modern cosmology, even a field theory of film. This dissipating focus is part of the point, for Pynchon leads us to the recognition that what he has rendered is not at all the simultaneous interactions of a field concept, but fallen, preterite versions of it may be all our cognitive consciousness can grasp. As the text plays with these transformations, we gradually realize that the point of the attempted returns to a single, unifying perspective is that there can be no true return, because we remain within the fragmented consciousness of modern analytical thought. More than any other writer in this study, Pynchon understands what it means to be caught in the cosmic web.

All these texts thus wrestle in some way with the implications of the field concept, from the first tentative imaginings of it in Lawrence to the exploration of the limits of imaging in Pynchon. It is in this rich diversity of strategies, the multiform ways the concept is transformed into literary form, that its importance for literature is found. For whatever stance these authors take toward the field concept, their encounter with it is affecting the shape of modern fiction.

Part I

MATHEMATICAL AND SCIENTIFIC MODELS

CHAPTER I

SPINNING THE WEB

Representative Field Theories and Their Implications

> . . . are the different styles of art an arbitrary product of the human mind? Here again we must not be misled by the Cartesian partition. The style arises out of the interplay between the world and ourselves, or more specifically, between the spirit of the time and the artist. The spirit of a time is probably a fact as objective as any fact in natural science, and this spirit brings out certain features in the world. . . . The artist tries in his work to make these features understandable, and in this attempt he is led to the forms of the style in which he works.
>
> Werner Heisenberg, *Physics and Philosophy*

THE SIGNIFICANCE OF the conceptual revolution in science derives less from the field models themselves than from their philosophical and epistemological implications. It is what they imply not only about the nature of the world, but about how one interacts with the world, that is important in understanding how the new view differs from the older, atomistic perspectives. One of the most important of these implications is that the Cartesian dichotomy between the *res cognitans* and the *res extensa,* the thinking mind and the physical object, is not absolute, but an arbitrary product of the human mind. Classical physics assumed that it was possible to make a rigorous separation between the observer and what she or he observes. Relativity theory and, in a different way, quantum mechanics require that the separation into an observer and a physical system be regarded as an arbitrary distinction entailing approximations that are not always negligible.

The breakdown of the Cartesian dichotomy also has methodological implications. When things are thought to exist "out there," separate and distinct from the observer, the world has already been divided into two parts. The next step is to subdivide it further by regarding the exterior world also as a collection of parts. The parts, because they are intrinsically separate and individual, can then be analyzed sequentially as individual units; this is of course how Aristotelian logic proceeds. As long as the world is conceived atomistically, this approach is appropriate and, at least in theory, exact to any desired degree of accuracy. But the field concept has the effect of revealing limitations in sequential analysis. These limitations are especially likely to appear when the whole is (or can be considered as) a part of itself.

For example, consider a set $\phi = \{a, b, c, d, \ldots\}$. In this example there is no problem in regarding the set ϕ as the whole, and each of the elements a, b, c, d as parts of that whole. But now imagine a set $a = \{a, b, c, d, \ldots\}$. From one perspective a is the whole itself, the entire set of elements enclosed within brackets. But from another perspective, a is a part of the whole, that is, one of the elements within the set. This problem is typical of paradoxes that arise from the field concept; it reveals an essential fallacy in the assumption that a whole can always be adequately defined as the sum of its parts. When classical, sequential analyses are applied to situations of this kind, paradoxes can become irresolvable antinomies.

I should like to turn now to more precise terminology and examine in some detail two examples in which the appearance of this kind of ambiguity proved to be decisive. In both cases, the paradoxes were revealed as a result of ambitious programs to extend the domain of classical analysis: in mathematics, the formalist program to prove that mathematics was free from contradiction; and in the philosophy of science, the positivist program to create an exact, objective language for science. These first examples are meant to convey a sense of how the generalizations I have been making about the field concept translate into specific examples from science. It is possible to see in them intimations of the complexities symbolized by the cosmic web.

In the early part of this century, the German mathematician David Hilbert suggested that it should be possible to prove that mathematics is free of contradictions by formalizing, one by one, the axiomatized theories of mathematics. Ernst Snapper, in a prize-winning article on the philosophical roots of mathematics,[1] explains that to "formalize" an axiomatized theory T means (confining ourselves to first-order examples) to choose a first-order language L so that all of the undefined terms that appear in the axioms of T can be expressed through parameters of L. It is then possible to express in L all the axioms, definitions, and theorems of T, as well as all the axioms of classical logic. In this approach, one manipulates the symbols of L by means of exact syntactical rules, without necessarily being concerned about the content of the symbols. The advantage of creating the language L is that L can then be studied as a mathematical object in itself, independent of the content of T. Hilbert hoped that a theory T could be proved free of contradiction by demonstrating that all of the allowable syntactical combinations of L were free of contradiction.

At the heart of this formalist program is the attempt to create a vantage point from which one could talk about mathematics as an object in a language that would not be contaminated with what it was one wished to prove. The Hilbert program rested on the assumption that it is possible to make a rigorous separation between the theory and the theory-as-object.

The hope that this strategy would succeed was shattered in 1931 with the publication of Kurt Gödel's paper, "Formally Undecidable Propositions in *Principia Mathematica* and Related Systems."[2] In this paper Gödel proved that for the mathematical system of the *Principia*, or more generally for any axiomatized theory with axioms strong enough so that arithmetic can be done in terms of them, the theory either will be inconsistent or will contain propositions whose truth cannot be demonstrated. Since inconsistencies are naturally to be avoided, mathematics finds itself impaled on the other horn of the dilemma; that is, it will

[1]Ernst Snapper, "The Three Crises in Mathematics: Logicism, Intuitionism and Formalism," *Mathematics Magazine*, 52 (September 1979), 207–216. This article won the coveted Allendoefer Prize in Mathematics for 1979.

[2]The German title of Gödel's 1931 paper is "Über formale unentscheidbare Sätze der *Principia Mathematica* und verwandter Systeme, I" in *Monatshefte für Mathematik und Physik*, 38 (1931), 173–198. For a fuller account of its implications for the formalist program, see Morris Kline, *Mathematics: The Loss of Certainty* (New York: Oxford University Press, 1980), pp. 245–306. Gödel's paper is reprinted in J. van Heijenoort, *From Frege to Gödel: A Source Book in Mathematical Logic* (Cambridge: Harvard University Press, 1967), pp. 596–616.

contain propositions that cannot unambiguously be proven to be either true or false.

Formally undecidable propositions had long been known and formulated through various paradoxes. One classic illustration is as follows. On the first side of a piece of paper write the words "The statement on the other side is true." Now turn the paper over and write "The statement on the other side is false." Let us consider first Side 1 asserting that Side 2 is true. If Side 2 is true, however, then Side 1 is false. But if Side 1 is false, then Side 2 is not true, in which case Side 1 *is* true. One can pursue this line of reasoning forever without being able to reach a conclusive answer. The two statements together involve what Douglas Hofstadter calls a "Strange Loop,"[3] a loop of reasoning that cannot be resolved because to accept either statement as true is to begin a loop which circles around to say that the same statement must be false. It is obvious such statements can be neither true nor false; they are inherently undecidable.

One way to analyze a Strange Loop is to consider it as a problem in self-reference. Each statement points to the other, and the other in turn points back, so that there is no independent vantage from which to evaluate either one. The Hilbert program had hoped to avoid this problem by separating the language L from the theory T. But this hope proved to be unfounded when Gödel demonstrated that it was possible to talk about number theory from within the theory itself. The problem of self-reference was thus revealed as unavoidable. Douglas Hofstadter explains:

> Gödel had the insight that a statement of number theory could be *about* a statement of number theory (possibly even itself), if only numbers could somehow stand for statements. The idea of a *code,* in other words, is at the heart of his construction. In the Gödel Code . . . numbers are made to stand for symbols and sequences of symbols. . . . And this coding trick enables statements of number theory to be understood on two different levels: as statements of number theory, and also as *statements about statements* of number theory.[4]

Using this method, Gödel was able to map statements about numbers

[3]Douglas R. Hofstadter, *Gödel, Escher, Bach: An Eternal Golden Braid* (New York: Basic Books, 1979), pp. 3–28 and *passim.*

[4]Hofstadter, p. 18. The italics are his.

into the number system itself. Recall that Hilbert's axiomatization attempted to create a strict separation between the theory and the theory-as-object. By making numbers stand for theoretical statements, Gödel circumvented this separation and thereby involved theoretical statements about numbers in paradoxes of self-reference, since numbers then became statements about numbers. These paradoxes led to the same sort of circular reasoning we saw earlier, with the result that the statements so involved could not be proven to be either true or false. Through this mapping procedure, Gödel was able to demonstrate that theories capable of embracing the theory of whole numbers cannot be both complete and consistent. If they are not inconsistent, then they will be incomplete, in the sense that they will contain statements which cannot be proven to be true under their axioms.

What happens if one takes the statements one cannot prove and converts them to axioms? (Axioms, of course, are unproven statements.) In this case one has generated a new theory, because the set of axioms has changed; and in this new theory, new statements will arise that cannot be proven within that system. If these new statements are in turn converted into axioms, still other statements will arise elsewhere in the system that cannot be proven under those axioms. The process is interminable.

The implication of Gödel's theorem, then, is that any theory that is not demonstrably false cannot be demonstrated to be completely true. Thus the program to prove all of mathematics true did not succeed. This does not necessarily mean that mathematics is false, of course—only that it cannot be proven true. The crux led Hermann Weyl to say that God must exist because mathematics is intuitively consistent, and the devil exists because it cannot be proven to be consistent. Whatever intuitive consistency one may grant mathematics, however, the inability to prove the truth of number theory is significant, for it reveals that even in mathematics, the most exact of the sciences, indeterminacy is inevitable.

Nor, it turns out, is this indeterminacy confined to axiomatic mathematics. It also appears in computation theory, in a problem that Martin Davis calls the Halting Problem.[5] The question that the Halting Prob-

[5]Martin Davis, "What Is a Computation?" in *Mathematics Today: Twelve Informal Essays,* ed. Lynn Sheen (New York and Berlin: Springer-Verlag, 1978), pp. 241–267.

lem asks is whether it is possible to determine in advance if a computer will be able to find a definite answer—that is, come to a halt—for any given problem.[6] The question has practical importance, for if it cannot be answered, one can suddenly find one's computer involved in a Strange Loop of its own, which consumes expensive computer time and, in extreme cases (as in the infamous "page fault" error), renders the program useless. The answer to the Halting Problem, Davis explains, is *no:* there will be some computations which cannot be proven in advance either to have a solution or not to have a solution, in much the same way that the Incompleteness Theorem says that there are some statements within number theory which cannot be proven to be true or false. In fact, Davis shows how Gödel's theorem (the Incompleteness Theorem) can be restated in terms of the Halting Problem, so that if the Halting Problem had a solution, the Incompleteness Theorem could not be true. Therefore, since the Incompleteness Theorem is true, the Halting Problem will not have a solution. The important point is that certain kinds of logical problems have no solution, not even using the most sophisticated computers imaginable. Davis makes this point explicitly: "Note that we are not saying simply that we don't know how to solve the problem or that the solution is difficult. We are saying: *there is no solution.*"[7]

What the Incompleteness Theorem does in mathematics, and what the Halting Problem does for the linear sequences of binary choices that comprise computer programs, is to imply that certain limitations in linear analysis are inescapable because of the problem of self-reference. It is because the tools for analysis are inseparable from what one wants to analyze that Strange Loops appear. In these examples, problems that cannot be solved through logical analyses appear as a result of considering both the tools for analysis, and the object to be analyzed, as part of the same "field." They illustrate one way in which the emergence of a field approach has revealed limits to classical logic.

In his introduction to *City of Words,* Tony Tanner explains that he has taken his title from the common thread he finds in contemporary fic-

[6]More technically, the problem asks whether there is a way to decide in advance if a universal program of the Church-Turing type will halt, given an initial input; Martin Davis explains these terms in detail, pp. 241–267.

[7]Davis, p. 255.

tion: its "foregrounding" of language.[8] Tanner's book has been influential not because it consistently maintains this focus—one reader complains that it degenerates into a "City of Themes"—but because, in suggesting that modern fiction is deeply concerned with the self-conscious use of language, Tanner has put his finger on a major characteristic of twentieth-century fiction. Modern readers are experiencing the same kind of situation that mathematicians experienced when Gödel's theorem burst upon the scene: the object for analysis (the text, number theory) refers self-referentially to that of which it is composed (language, statements within number theory). Like Gödel's theorem and the Halting Problem, modern fiction tends to place us within rather than outside the frame,[9] so that when we speak about it, we are speaking from within the picture that contains us. The resulting paradoxes have sparked important debates and theoretical work in literary criticism.

As we shall see in Chapter 6, Borges is well aware of this conjunction between mathematical and literary self-referentiality. In his story "The Aleph," Borges looks into a small sphere, "less than an inch in diameter," that contains everything in the earth, including another Aleph that contains within itself another earth . . . Borges's name for this sphere playfully alludes to Cantor set theory, for Georg Cantor chose to name his infinite sets "Alephs." The paradoxes that surfaced as a result of these infinite sets were instrumental in causing mathematicians to feel that it was necessary to axiomatize mathematics, and this in turn led to a realization that the paradoxes were not accidental but intrinsic to the structure of mathematics. As we explore these connections in Chapter 6, we shall see how, by transforming a scientific model into a literary sign, Borges makes it the basis for his distinctive narrative mode.

In the next example, the parallel between science and literature is even more apparent, for here the scientific debate was explicitly concerned with the nature of language. In the wake of the great successes of Newtonian mechanics, it seemed to many scientists that all physical phenomena would eventually yield to mechanical descriptions. Consid-

[8]Tony Tanner, *City of Words: American Fiction, 1950–1970* (New York: Harper & Row, 1971).

[9]An observation—and a phrase—that is the subject of Richard Pearce's "Enter the Frame," *TriQuarterly*, 30 (1974), 71–82.

erable attention was therefore devoted to refining scientific discourse so that it would establish unambiguously the link between this predictable reality and the theory that predicted it. The goal of the positivists was to "purify" language by removing from it anything that could not be empirically verified or logically demonstrated—in short, anything suspected of being "metaphysical." Statements that had "cognitive significance" were to be composed of three, and only three, kinds of terms: observational statements taken directly from experiment; theoretical terms; and logical terms indicating how the other two kinds of terms should be combined. Statements that did not fulfill these criteria did not possess "cognitive significance," or in plain words, were nonsense.[10] It was thought possible to extend the program beyond the experimental sciences into related fields such as the philosophy of science, and indeed to any field that proposed to engage in cognitively meaningful discourse. The attempt to reform scientific discourse is similar to Hilbert's mathematical program in that both strove for rigor by separating the object of discourse from the theory interpreting it. Like the Hilbert program, the positivist program failed when it was recognized that language creates a field that encompasses the observer as well as the observation.

In his history of the positivist program, Frederick Suppe recounts how the positivistic view of language, the heart of what he calls the "Received View" of scientific theories, was predominant in the philosophy of science through the early years of this century.[11] The "Received View" held that it was possible to distinguish unambiguously between theory and observation, and therefore possible to establish well-defined logical rules of correspondence between the two. The Received View came under increasing attack because the distinction between "observational terms" and "theoretical terms" could not be sustained as rigorous or complete. N. R. Hanson, for example, argued that what we see depends upon our cultural, scientific, and linguistic contexts.[12] Hanson pointed out that what the Received View had called "observational

[10]Positivists did recognize a genre called "emotive discourse," but whether this could be said to have meaning was considered problematic.

[11]Frederick Suppe, ed., *The Structure of Scientific Theories,* 2d ed. (Urbana: University of Illinois Press, 1977), pp. 6–56.

[12]N. R. Hanson, *Patterns of Discovery* (Cambridge: Cambridge University Press, 1958); Suppe summarizes Hanson's views at pp. 151–166.

terms" were in fact not sensory data per se, but sensory data as interpreted, at the very least, through an experimental apparatus that already had certain assumptions built into it, as well as through the unconscious perceptual sets of the observer. The positivist program was gradually yielding to the *Weltanschauungen* argument that observation was inherently theory-laden.

Thomas Kuhn took the argument further by suggesting that scientists, during their apprenticeships in their fields, absorbed a set of more or less unconscious assumptions about how science was "done." These assumptions, transmitted by learning model experiments or by mastering currently accepted theories, comprise the intuitive part of what Kuhn called the "paradigm" for that field.[13] Kuhn pointed out that there are always known facts that contradict accepted theories; but these will be ignored as long as the paradigm allows enough other data to be correlated satisfactorily. It is only when the paradigm begins to break down that anomalies will be noticed, or even reported. Only in this period of "revolutionary science," as Kuhn called the transition between paradigms, does an open-ended search for new kinds of facts come into play.

Michael Polanyi developed similar arguments in his analysis of "tacit knowledge," that is, knowledge which is in some sense known, but which cannot be formulated explicitly. It is the scientist's "tacit knowledge," Polanyi contends, that guides him to the *interesting* fact, the one datum or experiment out of thousands that will prove useful.[14] According to Polanyi, without this "tacit knowledge" science would degenerate into aimless forays or trivial experiments; it is the scientist's intuitive and nonverbal knowledge that gives direction to scientific inquiry and guides him toward significance.

Hanson, Kuhn, and Polanyi (along with others too numerous to mention here) have in common the belief that the distinction between "objective" facts and "subjective" reactions cannot be made in a comlete

[13]Thomas Kuhn, *The Structure of Scientific Revolutions,* 2d ed. (Chicago: University of Chicago Press, 1970). See also Kuhn's "Reflections on My Critics," in I. Lakatos and A. Musgrave, eds., *Criticism and the Growth of Knowledge* (Cambridge: Cambridge University Press, 1970), for Kuhn's refinement of the term "paradigm."

[14]A recapitulation of Polanyi's lengthy argument in *Personal Knowledge* (Chicago: University of Chicago Press, 1958) can be found in *Science, Faith and Society* (Chicago: University of Chicago Press, 1964).

or rigorous way.[15] They believe that what appear to be the "objective" facts of science are inextricably linked with important intuitive elements that are not susceptible to formal analysis or articulation. In this view, it is not possible to separate the observation from the scientist who observes. That the scientist's cultural and linguistic set helps determine what he or she sees implies that there is no way to create a language of observation that will not contain subjective elements. Thus self-referentiality has also entered in a crucial way into the question of whether it is possible to express scientific results in an objectively exact language. It has proven impossible to create such a language because the terms that comprise it already contain assumptions that cannot be validated independently of the language.

In *Zen and the Art of Motorcycle Maintenance* Robert Pirsig develops a similar argument by pointing out that any analytical hierarchy is created by an observer wielding a knife, even though the passive constructions of Aristotelian rhetoric work to conceal both the knife and, behind it, the observer who determines where the cuts will be made. In trying to find a rhetoric that will acknowledge that "*part* of the landscape, *inseparable* from it . . . is a figure in the middle of it, sorting sands into piles," Pirsig involves himself in the same paradoxes that the positivists encountered, for he himself is also "in the landscape," sorting into "piles" the different levels of narrative within his text. As we shall see in Chapter 3, it is when the narrator recognizes this paradox that the text comes to its climax and explodes into a series of contradictions that Pirsig cannot altogether control.

Concerning the two scientific examples discussed so far, the formalist program to reform mathematics and the positivist program to reform scientific discourse, a number of key issues have arisen—indeterminacy, self-referentiality, and the inability to make an unambiguous separation between subject and object—and they are linked by a common concern for the language. Recall that Gödel's theorem, the Halting Problem, and the *Weltanschauungen* analyses all emerged in response to programs that attempted to create a formally exact language. The question of how

[15]So successful have these *Weltanschauungen* analyses been that the trend in the philosophy of science now is alarm that we might lose sight of the logical and rational elements in science—see, for example, Suppe's "Afterword—1977," pp. 619–730. Although the field is still in disarray, it seems safe to say that any new view around which these positions might consolidate will have to incorporate at least some elements of the *Weltanschauungen* argument.

language is used, or, more accurately, how its use is perceived, is crucial because language mediates across the subject-object dichotomy. When this dichotomy is redefined in a the field concept, the perception of how language functions also changes.

Why language should play this key role will be apparent if we review the differences between the atomistic and field perspectives of language. In the atomistic view, the gap between subject and object is not "contaminated" by the circular paradoxes of self-referentiality because it is assumed that reality can be divided into separate, discrete components. Consequently, it is assumed that language can be used to define the relation between subject and object in a formally exact way. But the field concept assumes that these components are interconnected by means of a mediating field. When language is part of the mediating field (i.e., the means by which the relation between subject and object is described), it participates in the interconnection at the same time that it purports to describe it. To admit the field concept thus entails admitting that the self-referentiality of language is not accidental, but an essential consequence of speaking from within the field. As we have seen in a number of cases, when the atomistic approach failed it was because it proved to be impossible to create a language that would be free from problems of self-referentiality. Thus the shift from atomistic models to the field concept had the effect of bringing the self-referentiality of language into focus.

We are now in a position to develop further the parallels between modern literature and modern science. The modern novel emerged from exploring the Cartesian dichotomy in literary terms; or, to put the proposition in its more usual form, from exploring the relation between the teller and the tale. Modern physics developed from exploring the Cartesian dichotomy in scientific terms; or, to state it in its accustomed form, by exploring the relation between the observer and the observed system. Literary readers are well acquainted with the former assertion, scientific readers with the latter. What has not been sufficiently recognized by either is the isomorphism of the two propositions, and the resulting implication that both entail the self-referentiality of language. As self-referentiality of language is virtually the defining characteristic of post-modern criticism and texts, so is it also of post-Newtonian science. Whether the topic under discussion is Gödel's theorem or *Gravity's Rainbow,* self-referentiality is a crucial issue.

It also figures in an important way in the metaphor of the cosmic web, for it is what makes the web "sticky." This "stickiness" will become increasingly apparent as we turn to quantum mechanics and particle physics. First, however, it will be useful to understand in a more precise way how the assumptions of the older atomistic models, especially Newtonian mechanics, both reinforced and relied upon the Cartesian dichotomy, since it is the breakdown of the Cartesian dichotomy that brings the self-referentiality of language into focus as an important issue.

In classical mechanics, the physical world was considered to be composed of isolated objects separated from one another in an empty space that was rigid and unchanging, with a universal "now" pervading all space at any given moment. Because time was handled as though it consisted of a succession of universal moments, there was never any ambiguity about the order of events. Hence causality could be unidirectional and absolute. Moreover, the kind of causality predicted by the equations of classical mechanics was thought to have been laid down at the creation of the world as immutable principle. Albert Einstein recounts how the generations of physicists preceding him believed that "God created Newton's laws of motion together with the necessary masses and forces . . . everything beyond this follows from the development of appropriate mathematical models by means of deduction."[16] Since these laws were unchanging, they held good for the indefinite future. It was in theory enough to know the initial set of conditions and Newton's equations of motion to predict any future state, assuming only sufficient intellect (or computer space) to do the calculations. The great French mathematician Pierre Laplace imagined "an intellect which at a given instant knew all the forces acting in nature, and the position of all things of which the world consists"; this vast intellect could then "embrace in the same formula the motions of the greatest bodies in the universe and those of the slightest atoms; nothing would be uncertain for it, and the future, like the past, would be present to its eyes."[17] In the classical model, the emphasis thus fell on well-defined interactions that could be exactly predicted by the Newtonian equations

[16]Albert Einstein, *Autobiographical Notes,* trans. and ed. Paul Arthur Schilpp (La Salle, Ill.: Court, 1979), p. 19.

[17]Quoted in Milic Čapek, *The Philosophical Impact of Contemporary Physics* (Princeton: D. Van Nostrand, 1961), p. 122.

of motion and projected infinitely far into the future. The equations themselves were considered immutable and complete, not susceptible to further change or modification.

These assumptions also had important methodological implications. Because interactions were unidirectional, the dominant mode by which systems were related to one another, and hence the dominant mode of analysis, were causal. Because the physical world consisted of discrete bodies separated in space, analysis of systems could be carried out through interlocking series of discrete logical steps. Because systems were already inherently discrete, there was no problem in separating the observer from what he observes. And finally, because the physical world existed "out there," independent of the observer, it was determinate and infinitely knowable. There were no theoretical limits to how much the rational mind could understand about the physical world because the mind, in understanding physical reality, did not have simultaneously to understand itself.[18]

All of these assumptions were fundamentally questioned, and finally overthrown, by developments emerging from two papers that Albert Einstein published in 1905. One, drawing on Max Planck's suggestion that light was quantized, was instrumental in the creation of quantum mechanics; the second set forth the Special Theory of Relativity. With these two seminal papers, the new physics was launched. In a little over a decade Einstein would extend his conclusions to the General Theory of Relativity. Meanwhile, intense attention was being devoted to quantum phenomena, and by 1927 the mathematical formalism of quantum mechanics was essentially complete. With the formalism and theories in place, the debate on what they meant began in earnest. What became increasingly clear throughout the subsequent decades was that the new scientific models implied not only a new physics, but a new world view.

Before physicists became concerned about such questions as self-referentiality, indeterminacy, and the lack of a rigid separation between

[18]For a fuller and more precise explication of the model of reality implied by Newtonian mechanics, see Clifford Hooker's excellent analysis in "The Nature of Quantum Mechanical Reality: Einstein versus Bohr," in *Problems and Paradoxes: The Philosophical Challenge of the Quantum Domain*, ed. Robert G. Colodny (Pittsburgh: University of Pittsburgh Press, 1972), pp. 69–72. Hooker concludes his analysis of the classical model of reality with this observation: "The general conception of the physical world conveyed in the preceding statements will no doubt be familiar to the reader. *It is a measure of the revolution brought about by the advent of the quantum theory that every one of these claims has been challenged*" (p. 72, italics his).

subject and object, they encountered the startling ways in which the field concept transformed traditional views of time and space. With characteristic generosity Einstein, in a tribute written on the hundredth anniversary of James Clerk Maxwell's birth, attributes to him this revolutionary change in notions of physical reality. Maxwell is remembered for his work in developing a field theory that united magnetism and electricity into the single entity that is now called the electromagnetic field. Before Maxwell, Einstein remarks, "people conceived of physical reality—in so far as it is supposed to represent events in nature—as material points, whose changes consist exclusively of motions, which are subject to total differential equations. After Maxwell they conceived physical reality as represented by continuous fields, which are subject to partial differential equations."[19] Maxwell had established the notion of a field as a concept equal in explanatory power to the Newtonian idea of material points when he showed how electromagnetic phenomena (including light) could be represented through a system of differential equations. Even a writer like D. H. Lawrence, who understood little of the mathematics, grasped the essence of this change and fashioned a literary model of it in the "polarities" and "fields" that we shall encounter in Chapter 4. Lawrence understood also that Einstein was connected with this transformation and that Einstein, even more than Maxwell, was "knocking that eternal axis out of the universe." In this premonition Lawrence was correct, for it was Einstein who, in relativity theory, gave Maxwell's classical notion of a field its most powerful expression.

So much nonsense has been written on the implications of relativity that one can sympathize with Martin Gardner's abrupt dismissal of the topic in his popular book on relativity. "If the reader wonders why the book contains no chapter on the philosophical consequences of relativity," Gardner remarks, "it is because I am firmly persuaded that in the ordinary sense of the word 'philosophical," relativity *has* no consequences."[20] Gardner's assertion is an overstatement, for as he goes on to admit, relativity theory does have important epistemological implica-

[19]Albert Einstein, "Maxwell's Influence on the Evolution of the Idea of Physical Reality," *Ideas and Opinions* (New York: Dell, 1954), p. 262.

[20]Martin Gardner, *The Relativity Explosion*, rev. ed. (New York: Vintage Books, 1976), p. x.

tions. But it is necessary to sort out what relativity does and does not imply.

Relativity does *not* imply that "everything is relative." Indeed, before he settled on "relativity," Einstein had considered calling his hypothesis the "Theory of Invariance." In his *Autobiographical Notes* Einstein says that he believes scientific theories should possess what he calls "logical simplicity," that is, that their fundamental postulates should not be the result of arbitrary restrictions but should flow naturally from the initial conception.[21] What Einstein found "particularly ugly" about Newtonian mechanics was that it gave special priority to stationary or nonaccelerating systems over all other kinds of rigid systems, without it being obvious why this should be so.[22] Similarly, it had been hypothesized that it was possible to define absolute motion by regarding all motion as taking place within an "ether," an invisible and virtually undetectable medium that was supposed to permeate space. In retrospect it is evident that these restrictions were necessary to preserve congruity with everyday experience. As Werner Heisenberg points out, the concepts of classical physics—mass, velocity, momentum, force—are simply the experiences of everyday life cast into more exact and rigorous terms.[23]

Relativity theory, by contrast, derives many results that are startlingly at odds with everyday experience. Rather than beginning with "common sense," Einstein's thought was guided by a search for harmony among fundamental principles. It is this, rather than its extraordinary predictions, that struck Cornelius Lanczos, a physicist of Einstein's generation, as the most revolutionary aspect of relativity theory. Einstein saw science in a new light, Lanczos comments. "To him science did not mean the primacy of the experiment or the primacy of the theory, but the primacy of a deep reverence for the all-embracing lawfulness which manifests itself in the universe."[24]

Einstein's allegiance to fundamental principle can be seen in his account of how he arrived at the Special Theory. When he was sixteen, Einstein tried to imagine how a light wave would look to someone

[21]Einstein, *Autobiographical Notes,* p. 21.

[22]Ibid., p. 25.

[23]Werner Heisenberg, *Physics and Philosophy: The Revolution in Modern Science* (New York: Harper & Row, 1958), p. 56.

[24]Cornelius Lanczos, *Albert Einstein and the Cosmic World Order* (New York: Interscience Publishers, 1965), p. 112.

traveling at the speed of light.[25] He decided that to such an observer, the light beam would appear as a standing wave, oscillating back and forth without forward movement. This result puzzled Einstein not only because it was contradicted by Maxwell's equations, which implied that nonpropagating light was impossible, but more fundamentally because it implied that phenomena can appear different from different vantage points. Einstein decided that if he had to choose between the laws of physics being universal or phenomena appearing invariant, he would choose the laws of physics. In the Special Theory, Einstein begins by assuming that the laws of physics should not depend on whether one is at rest or in uniform motion. He also assumes that the velocity of light in a vacuum is constant, regardless of the motion of its source. In order to preserve these invariances, Einstein reasoned that motion could only be defined relative to some arbitrarily chosen reference frame. With this reasoning, Einstein arrived at the now-familiar predictions that measurements of time, mass, and length are not absolute quantities but subject to change, depending on the reference frame from which they are made.[26] Paradoxically, these quantities are made relative so that others may become absolute. The primary absolute is that the laws of physics remain invariant for any rigid system in uniform motion.

A more sweeping absolute emerges from the interpretation that Hermann Minkowski, the Polish mathematician, gave to the interdependence of time and space in the Special Theory. As he set forth his interpretation before an assembly of colleagues, Minkowski predicted, "Henceforth space by itself, and time by itself, are doomed to fade away into mere shadows, and only a kind of union of the two will preserve an independent reality."[27] In the Minkowski interpretation, time and space are combined into the four-dimensional matrix of "spacetime." It is when this four-dimensional matrix is projected into the three dimensions of traditional Cartesian space that different observers can disagree about what happened. If, however, time is added as a "fourth dimen-

[25]Recounted in Fritjof Capra, *The Tao of Physics: An Exploration of the Parallels between Modern Physics and Eastern Mysticism* (New York: Bantam Books, 1977), pp. 153ff.

[26]It should be emphasized that the relativity of these quantities is not merely a perceptual ambiguity in the observer. The most sensitive instruments (for example, nuclear decay clocks) will record a time that is not absolute, but relative to the reference frame to which they are attached. Bertrand Russell makes this point with special clarity in *The ABC of Relativity,* rev. ed. (Fair Lawn, N.J.: Essential Books, 1958), p. 133.

[27]Hermann Minkowski in *The Principle of Relativity* by H. A. Lorentz, A. Einstein, H. Minkowski, and H. Weyl (New York: Dover, 1923), p. 75.

sion," the resulting (four-dimensional) description will be the same for all observers. By thus expanding the traditional three-dimensional Cartesian space into a four-dimensional matrix, invariance is achieved. In E. F. Taylor and J. A. Wheeler's words, "Space is different for different observers. Time is different for different observers. Spacetime is the same for everyone."[28] The absolute time and absolute space of Newtonian physics have thus given way to a new absolute composed of both time and space.

In the General Theory, Einstein extended his conclusions by postulating that the laws of physics are invariant not only for bodies in uniform motion but also for bodies in accelerating motion, so that the long-recognized equivalence of gravitational mass and inertial mass (the "weight" an accelerating object will assume in space, as a result of inertial resistance to the acceleration) is established theoretically. Thus not only the choice of reference frame became arbitrary, but also the type of motion, for accelerating systems are treated in the General Theory with the same equations as nonaccelerating or stationary systems. As a result, a radically different view of spacetime emerged. In the General Theory, gravitation is seen not as some mysterious force that mass exerts over distance, but as a result of the nature of spacetime itself. Einstein suggested that we should think of spacetime as being curved around large masses, and that it is this curvature which accounts for gravitational phenomena. Spacetime, in this view, is not an empty container for mass. Rather it exists, and is given its characteristic structure, because of the distribution of mass. Indeed it cannot, properly speaking, be considered apart from mass. Whereas the Special Theory joined space and time into the single field of spacetime, the General Theory further correlated spacetime and mass, regarding gravitation as a physical expression of the interaction between them.

In both the Special and General Theories, then, Einstein arrived at a view of physical reality that transformed the isolated entities of Newtonian mechanics into unified, mutually interacting systems. Instead of seeing time as a series of independent and omnipresent moments, Einstein conceived of it as inextricably linked with space to form the four dimensions of spacetime; instead of thinking of space as a rigid container, Einstein postulated that it took its structure from matter; instead

[28]The passage is from *Spacetime Physics,* quoted in Gardner, p. 101.

of seeing energy and matter as fundamentally separate and inconvertible, Einstein showed that they are essentially equivalent and potentially interconvertible. In all these results, relativity theory had the effect of transforming isolated parts into an interconnected whole. In seeing fundamental interconnections between entities that had been discrete quantities in classical physics, Einstein helped to prepare the way for a field concept of reality—whose more radical implications, however, he was to resist for the rest of his life. Einstein deeply believed in causality, in an objective world that exists independently of human perception, and in the universal truth of scientific law. As we have seen, all of these notions come into question when the field concept is expanded to include the language of observation, whether natural or scientific. With quantum mechanics, especially as interpreted by Niels Bohr, this expansion took place within physics itself.

Meanwhile, even Einstein's classical formulations were disquieting to many of his contemporaries, because they involved a new way of looking at the world as an interconnected, mutually interactive unity. Cornelius Lanczos recounts how a colleague walked out of an early seminar on relativity in disgust, remarking "I am a physicist, not a philosopher." Lanczos himself admits, "To get used to this much more abstract way of thinking [necessitated by relativity theory] was not easy."[29] But he also argues that the "gradual abstractization of our primitive concepts" that "may appear on the surface as a loss" is more than offset by the gain. "We admit the loss of simplicity," Lanczos remarks, "but we are willing to pay the price for the tremendous advance in *unity*."[30] Einstein himself saw the advance in unity as the decisive factor. In a lecture at Princeton University in 1921, Einstein commented: "The possibility of explaining the numerical equality of inertia and gravitation by the unity of their nature gives to the general theory of relativity, according to my conviction, such a superiority over the conceptions of classical mechanics, that all the difficulties encountered must be considered as small in comparison."[31]

But this is unity of a very special kind. If relativity asserts that apparently different phenomena follow the same general laws, it emphasizes that our particular experience of those phenomena is not especially

[29]Lanczos, p. 109.
[30]Ibid., p. 110.
[31]Quoted in Gardner, p. 83.

privileged. The angle from which we view the universe is only one among many, no more (or less) valid than any other. Relativity theory permits a more general formulation of the laws of physics; but at the same time any perspective from which we might actually view the world is made partial and contingent.[32]

Relativity, then, contains two fundamental and related implications that were to be absorbed into the field concept: first, that the world is an interconnected whole, so that the dichotomies of space and time, matter and energy, gravity and inertia, become nothing more than different aspects of the same phenomena; and second, that there is no such thing as observing this interactive whole from a frame of reference removed from it. Relativity implies that we cannot observe the universe from an Olympian perspective. Necessarily and irrevocably we are within it, part of the cosmic web.

It is precisely this relativity of viewpoint that Nabokov resists in *Ada*, though he is eager to explore the related proposition that relativistic time is not susceptible to uniform measurement. Nabokov's treatment of relativity theory in *Ada* is as selective as his narrator's, who renounces the "space-tainted, space-parasited time . . . of relativist literature" while still arguing that the measurement of "real" time is variable. The implicit strategy behind this selectivity is at the center of the discussion in Chapter 5, for it reveals how an artist can shape a model for his own ends, and how this shaping can be at once scientifically incoherent and artistically powerful. For this purpose *Ada* is a key text, because the ambiguities and tensions between what Nabokov borrows from relativity theory and what he rejects are central to the novel's artistic strategies.

If Einstein is the father of relativity theory, he is the disapproving stepfather to quantum mechanics, the discipline sparked by his other

[32]The partiality of our own perspective should not be confused with the absolute spacetime projected through the Minkowski diagrams. David Bohm, in *The Special Theory of Relativity,* comments that when viewing Minkowski diagrams, "almost unconsciously, one is led to adopt the point of view of an observer who is, as it were, standing outside of space and time . . . surveying the whole cosmos from beginning to end" (p. 173). But as Bohm points out, the feeling is an illusion. Any human observer is necessarily in space and time, and so always in fact occupies a point *within* the diagram. Similarly, the timeless nature of the Minkowski diagrams (timeless in the sense that time is encompassed in the spacetime matrix) should not lead us to think that this reality already preexists. Relativity theory, insofar as it says anything about the future, is fully consonant with seeing it as a Becoming rather than a Being. Milic Čapek discusses this point extensively (pp. 214–243).

early paper. Relativity theory established a connection between the observer and the observation; in quantum mechanics, they are wed into an indissoluble whole. Despite profound philosophical differences, quantum mechanics is like relativity theory in that it joins together concepts that were quite distinct in classical physics—particles and waves. In classical physics, matter consisted of discrete particles that were localized in space and that had a definite trajectory through time. Electromagnetic waves, on the contrary, propagated through space much as sound waves do through the air, and hence were nonlocalized and capable of interference phenomena. But the two-slit experiment on electrons showed that in some circumstances electrons displayed interference phenomena, while the photoelectric effect demonstrated that light can act as if it were composed of a stream of particles. Electrons thus sometimes act like waves, while light sometimes acts like particles. This ambiguity is formalized in the Heisenberg Uncertainty Relation, which is a mathematical expression of the limits within which a particle can be localized. The Uncertainty Relation is concerned with how precisely the position and momentum of a particle can be known simultaneously.[33] The more sharply the one value is determined, the more diffuse the other becomes; the product of the uncertainties in momentum and position cannot be less than a universal constant known as Planck's constant.

The wave/particle duality is further expressed in the mathematical functions that quantum mechanics uses to describe "particle" behavior. These are wave functions of finite length, or "wave packets." Because particle density varies in accord with the wave function, the expressions can be interpreted as the probability that the "particle" will be at a given location. But the particle in quantum mechanics should not be thought of as a particle in the classical sense. It is not a discrete entity localized in space, but a "probability wave," the probability expressing the particle's "tendency to exist" at a given point.

One of the earliest physical interpretations of the Uncertainty Relation, still given in many textbooks,[34] came from Heisenberg's "thought

[33]The Uncertainty Principle can be extended to any two conjugate variables that do not commute in quantum theory, for example time and energy.

[34]See for example Richard P. Feynman, Robert B. Leighton, and Matthew Sands, *The Feynman Lectures on Physics,* vol. 3 (Reading, Mass.: Addison-Wesley, 1963–1965), pp. 1.1–1.8.

experiment" with a gamma-ray microscope. By closely analyzing how a very small particle—for example, an electron—is "seen," Heisenberg showed that the quantum of light used to observe the electron is sufficient to change the particle's momentum. Therefore, by the time the image is reflected back to the microscope lens, the particle is no longer following the same path it was because observing it has also disturbed it. If light of a lower frequency (and hence less energy) is used so as to disturb the particle less, the longer wavelength means that the reflected image will not be as sharply localized. Thus, as the momentum becomes more precise because the particle is not disturbed as much, the position measurement grows less precise. The more precisely the momentum is known, the less precisely the position can be known.

Heisenberg's analysis had a revolutionary impact because it made clear that the indeterminacy set forth by the Uncertainty Relation is not just a result of limitations in the measuring instruments, but fundamental to the process of measurement itself. It implies that there is no way to measure a system without interacting with it, and no way to interact with it without disturbing it. The observer and the system, or as Heisenberg occasionally said, the subject and object, are thus seen as an inseparable whole that cannot be subdivided without introducing the indeterminacy specified by the Uncertainty Relation.

Its continued popularity notwithstanding, Heisenberg's "disturbance" language raises perplexing questions. Are we to understand, for example, that the particle had a determinate value *before* it was measured? Heisenberg sometimes answered by asserting that it is not meaningful to talk about a reality that by definition could never be measured; whether in fact there is a reality "out there," prior to measurement, in this view is irrelevant. Under the influence of Niels Bohr, Heisenberg gradually came to the view that the process of measurement in some way *determines* the values, brings into actuality what was before a potentiality. In *Physics and Philosophy,* Heisenberg argues that in any physical experiment, "what we observe is not nature in itself but nature exposed to our method of questioning."[35] He therefore suggests that we should replace the concept of an objective reality as a thing-in-itself (*Ding an Sich*) with the Aristotelian idea of a "potentia." This "potentia" is "something standing in the middle between the idea of an event

[35]Heisenberg, *Physics and Philosophy,* p. 58.

and the actual event, a strange kind of physical reality just in the middle between possibility and reality."[36] According to Heisenberg, through interaction with the observing system the potential is partly transformed into actuality, though its quality as a "potentia" is never completely lost; some indeterminacy (implied by the Uncertainty Relation) remains.

Heisenberg frequently speaks of Bohr's interpretation of the Uncertainty Relation as if it were synonymous with his own. Partly because of this, the two together have come to comprise what is usually called the "Copenhagen interpretation." But as Clifford Hooker observes in his excellent analysis of Bohr's philosophy, Bohr's position was really quite different. Bohr never endorsed the "disturbance" concept of Heisenberg. Rather, in his view, the Uncertainty Relation was deeply bound up with the limitations of language. Bohr's long-time colleague Aage Petersen recounts how Bohr loved to repeat that "we are suspended in language."[37] In Bohr's view, the question of language is crucial; and it is in his philosophy that the connecting links between a field view of language and the field concepts of quantum mechanics are clearest.

According to Bohr, we define matter and energy through the terms of classical physics as either particles or waves; but because they are neither one nor the other, either description will be incomplete in precisely the way laid down by the Uncertainty Relation. What the Uncertainty Relation implies is a "quantum of action," a term that Bohr took over (its usual meaning is the numerical value of Planck's constant) to denote an area within which no further distinction between the observer and system is possible. "The fundamental postulate of the indivisibility of the quantum of action," Bohr writes in an early essay, ". . . forces us to adopt a new mode of description designated as *complementary* in the sense that any given application of classical concepts precludes the simultaneous use of other classical concepts which in a different connection are equally necessary for the elucidation of the phenomena."[38] In short, if we describe the phenomenon as a particle, we miss its wavelike properties; if we describe it as a wave, we miss its corpuscular properties.

[36] Ibid., p. 41.

[37] Aage Petersen, "The Philosophy of Niels Bohr," *Bulletin of the Atomic Scientists*, 19 (September 1963), 10.

[38] Quoted in Hooker, p. 138. The italics are Bohr's; Hooker's italics have been omitted.

For our purposes, the most interesting aspect of this argument is Bohr's reason why we cannot simply abandon the classical terms and seek others. It is here that Bohr's idea of being suspended in language enters. The classical concepts, Bohr felt, evolved as a consequence of our experience in the world; they reflect the essential distinction between subject and object that is the absolute prerequisite for the process of observation to begin. From the division into subject and object, Bohr writes, "follows . . . the meaning of every concept, or rather every word, the meaning depending upon our arbitrary choice of viewpoint."[39]

The very act of speaking, Bohr felt, evolved from the distinction between the subject and object. To speak is to speak from a position that is defined as separate and distinct from that which is spoken about. Language thus implies a viewpoint, a specific place at which the subject-object split is made. But because of the Uncertainty Relation, this viewpoint will always result in an incomplete and partial description. To complete the description, another viewpoint is necessary which makes the subject-object split in a different place. But these viewpoints will be mutually exclusive, because the subject-object split can only be made in one place at a time. Hence no matter which viewpoint is chosen, there will always be aspects of reality that can only be understood from another, mutually exclusive viewpoint. To switch to that new viewpoint will render indistinct and hence indeterminate aspects that may have been clear in the former viewpoint. Consequently Bohr affirms that "we must, in general, be prepared to accept the fact that a complete elucidation of one and the same object may require diverse points of view which defy a unique description."[40] The classical concepts cannot simply be abandoned, because any concept whatever—that is, any definition of reality that is external to us—will have the same built-in limitations of viewpoint.

Although Bohr does not rely on linguistics in making this argument, it is possible to recapitulate his reasoning in these terms, through a consideration of the deep structure of Indo-European languages. To make a well-formed utterance in English, for example, is implicitly to acknowledge a structural division between actee and actant, as well as

[39]Ibid., p. 141.
[40]Ibid.

the temporal progression implicit in verb tenses. Thus not only is a speaking subject posited in opposition to an "outside" world, but that relationship is further defined as occurring at a particular place in time and space. Hence Bohr's point—that to speak requires a subject-object dichotomy—is true not only in the general sense that to speak is to assume a separation between the speaker and the object of speech, but also in the more specific linguistic sense that to speak is to use a linguistic structure built on such distinctions.

It is this sense of being trapped inside the conceptualizations of language that, more than anything else, keeps surfacing in Pynchon's *Gravity's Rainbow* as the fatal barrier separating humanity from full participation in a holistic reality. Though Roger Mexico can argue for a world based on probability, neither one nor zero but somewhere in-between, Pynchon shows human cognition as fundamentally bound up with the binary distinctions characteristic of black-and-white films, inorganic chemical reactions, and the human neural system. As Pynchon explores the dependence of cognition on breaking a unified field into separate and isolated components, he mourns for the holistic, nonfragmented reality that he imagines other species can sense. The inevitable end of our relentless forcing of a holistic field into atomistic perspectives, Pynchon suggests, will be the destruction of a humanity which can never be "simply here, simply alive."

For Bohr, the fact that we remain "suspended in language" does not mean that we cannot make progress; he would therefore be unwilling to subscribe to Pynchon's fatalistic view. According to Bohr, we progressively refine our viewpoints not by attempting the impossible, that is, observing without a viewpoint, but by recognizing the ways in which our description of reality depends on the viewpoint we have chosen. "The development of physics has taught us that . . . even the most elementary concepts . . . [are] based on assumptions initially unnoticed," Bohr writes. When an "explicit consideration" of these concepts is undertaken, we "obtain a classification of more extended domains of experience." These more extended domains will in turn be underlaid by other concepts containing "unrecognized presuppositions," the examination of which will in turn lead to a still more general description.[41] Thus progress is made not by ignoring or underplaying

[41]Ibid., p. 139.

limitations of viewpoint, but by systematically examining and exploiting them.

It would be possible to write the history of the modern novel using similar terms, starting from a Jamesian theory of point of view and progressing to post-modern literature in which the assumption that there is a "point" from which to "view" is called into question. Shifting viewpoints that are mutually exclusive but all in some sense true; experiments that involve making the subject-object split in different places; the radical questioning of what it means to be "objective"—these are familiar to literary readers as the central issues of modernism, just as the problem of self-referentiality is the central issue of post-modernism. If we follow Bohr's advice (and Pynchon's example), the next step is to examine the underlying assumptions behind these literary strategies, thereby preparing the way for yet another enlargement of our understanding.

Before turning to this task in the remaining chapters, however, we will find it useful to look one last time at the scientific models, now concentrating not on what they have accomplished, but on what they have failed to accomplish. If the isomorphism between the scientific models and literary strategies holds, these limitations will have something to tell us about related limitations in the literary strategies.

Throughout this chapter, two themes have been implicit, and I should like now to state them explicitly. One is the extraordinary vision of unity inherent in the field concept of reality; the second is the extreme difficulty of translating this intuitive vision into an articulated model. The difficulties of constructing a conceptually coherent model are apparent in the uneasy alliance of relativity theory and quantum mechanics. Why is the alliance uneasy? Because the thrust of quantum mechanics, as we have seen, is to render indeterminacy inherent, while the thrust of relativity theory is to extend the determinacy of Newtonian physics into the progressively larger unifications made possible by Einstein's assumptions of invariance. Thus while quantum mechanics is probabilistic rather than causal, nonlocal rather than local, in relativity theory Newton's gravitational "action-at-a-distance" is replaced with strictly local action. In relativity theory force is considered to be mediated by means of an underlying field, and the field itself is considered to be mediated through the exchange of particles. Hence the existence of gravity, for example, implies that there should also be "gravitons" and "gravity waves" (though no generally accepted detection of them has yet been made). As a result, relativity theory, in contrast to quantum mechanics, is determinate rather than indeterminate, a theory of local action instead of action-at-a-distance.

The dilemma for modern physicists is that both relativity theory and quantum mechanics have proven so successful within their respective spheres of applicability that it is highly unlikely either will be abandoned; moreover, *both* are clearly necessary when dealing with atomic phenomena. Though no entirely satisfactory way to combine the two has yet been found, the difficulties are mostly in combining quantum mechanics with general relativity; the blend with special relativity has been very successful, and quantum field theory is now well established. But because the conceptual differences between the relativistic and quantum theoretics persist, various other models have gained a hearing in the scientific community, among them "hidden variable" theories. These theories, regarded as untenable by many physicists, show how very different models, some of them conceptually very strange, can emerge from a view of reality on which there is general consensus.

Hidden variable theories postulate that in some way that is not clearly understood, "certain dynamical variables" are affected when two particles interact. Thus they assume that the unknowable area covered by Bohr's "quantum of action" is in effect controlled by the "hidden variables," whose presence we may infer even though they are "hidden" from sight. In general, hidden variable theories were an attempt to restore determinism and causality to quantum mechanics by postulating a causal mechanism operating within the area of uncertainty.

The efforts of the hidden variable theorists took a dramatically different turn, however, when J. S. Bell, in what is usually called "Bell's Theorem," showed that a hidden variable theory cannot reproduce all of the statistical predictions of quantum mechanics unless it gives up the assumption of local action. As a result, some hidden variable theories adopted a non-locality assumption that, in the words of Max Jammer, endowed them "with features that seemed to belong to magic rather than physics."[42] They assume, for example, that a connection between two particles can obtain even though they are widely separated in space.

[42]Max Jammer, *The Philosophy of Quantum Mechanics* (New York: John Wiley, 1974), p. 311.

In this assumption particle A, for example, could be influenced by the kind of measurement performed on particle B if A and B had at some previous moment been in touch, even if at the time of the measurement A and B are widely separated and have no further interaction. The two systems are thus supposed to be united in what Jammer characterizes as a "mysterious conspiracy." "Even to many nonconformists," Jammer concludes, "Bohr's complementarity interpretation seemed to be less bizarre."[43]

In contrast to the intuitive implausibility of the model, however, is the shared vision of what a field view of reality entails. David Bohm, one of the leading hidden variable theorists, emphasizes that what he calls the "implicate order" implicit in hidden variable theory is in harmony with both relativity theory and quantum mechanics.[44] According to Bohm, relativity theory and quantum mechanics have in common "the notion of unbroken wholeness"; "if relativity were able to explain matter, it would say that it would be all one form—a field—all merging into one whole. Quantum mechanics would say the same thing for a different reason, because the indivisible quantum links of everything with everything imply that nothing can be separated." Bohm therefore suggests the emergence of an implicate order "which will be suitable for this unbroken wholeness."[45] In the implicate order, "each part . . . contains the whole in some sense. The whole is folded into each part." In this view "points are not the fundamental notion any more as in the Cartesian system. Rather, what is fundamental is some region which contains, in some sense, the order of the whole."[46]

The contrast between the simplicity of the vision and the difficulty of the model is also apparent in many mainstream theories. Einstein, although he did not succeed in formulating a unified field theory that would unite relativity and quantum mechanics, nevertheless had a clear vision of what it would imply. In such a theory matter would be regarded as "being constituted by the regions of space in which the field is extremely intense. . . . There is no place in this new kind of physics

[43]Ibid., p. 312.
[44]David Bohm, "The Implicate Order: A New Order for Physics," *Process Studies*, (1978), 73–102; a fuller treatment of these ideas can be found in *Wholeness and the Implicate Order* (London and Boston: Routledge & Kegan Paul, 1980).
[45]Bohm, "The Implicate Order," p. 90.
[46]Ibid., p. 91.

for both the field and matter *for the field is the only reality*."[47] Similarly, the prominent mathematician and physicist Hermann Weyl wrote years ago that the electron should be considered as "merely a small domain of the electrical field within which the field strength assumes enormously high values."[48]

Clifford Hooker, in 1972, suggested that the key to reconciling this shared vision and competing models may lie in an essential change of perspective. "The general presupposition behind fundamental particle theory," Hooker writes, "is that there is a subatomic structure to physical reality, that just as macro bodies actually consist of atoms, so atoms actually consist of fundamental particles, and so on down." As Hooker points out, this view implies that the theories will assume a certain form, "where particles in hierarchy level n are seen as structured swarms of particles of level n-1." But suppose, Hooker continues, that "the so-called subatomic world was only nature's way of responding to high energy attacks. Suppose, for example, that the world were really continuous and the manner of its apparent breaking up was much more like the water droplets ejected as a stone strikes the surface." In this case the proliferation of fundamental particles is "best understood from the top down," as "characteristic denizens of our machines only," rather than as "revealing a pre-existing physical structure to be discovered." In words that David Bohm would echo six years later, this view of atomic phenomena in which it is seen "from the top down" would "turn theorizing, and experimenting, on its head."[49] The turn of thought, from a view that sees the essence in the smallest indivisible part to a view that sees the essence as an indivisible whole, is clear. What remains unclear is whether it can ever be adequately expressed in an articulated model.[50]

[47]Quoted in Čapek, p. 319.

[48]Hermann Weyl, *Philosophy of Mathematics and Natural Science* (Princeton: Princeton University Press, 1949), p. 171.

[49]Hooker, p. 179.

[50]For a recent survey of where the matter stands now, see Gerard t'Hooft, "Gauge Theories of the Forces between Elementary Particles," *Scientific American*, 243 (June 1980), 104–137. T'Hooft reports that it now appears possible to represent all four kinds of interactions between elementary particles by the same general kind of theory. This implies that it may one day be possible to unite all four interactions under a common theoretical framework, resulting in the unified field theory of Einstein's dream. Although no such theory has yet been found, a step in this direction was taken with quantum electrodynamics, which allows the wave/particle duality to be correlated with electromagnetic fields. But the problems encountered testify to the difficulties of conceptualizing reality as a unified field. T'Hooft recounts how the search for a workable model led to such

What our survey of the field concept in various scientific models has shown is that the problem of articulation is intrinsic to this view of reality, whether the language involved is the binary sequence of computer programs, the "wave-packet" equations of quantum mechanics, or one of the syntactically linear natural languages in which scientists attempt to come to grips with the philosophical implications of their models. Because the task of articulation requires that a vision of a dynamic, mutually interacting field be represented through a medium that is inherently linear, fragmented, and unidirectional, the novelist's concern with language will have much in common with these scientific concerns. The strain of trying to capture the idea of a holistic field in an articulated medium will thus be as apparent—and as interesting—in the literary chapters as it has been in this chapter on scientific models. The authors to whom we now turn have their own perspective and insights to bring to this problem. Whereas the scientific theories are created through the attempt to express the field view in rigorously exact models, the literary strategies are forged by the desire to find a form, and a language, adequate to interpret its human meaning.

---

expediencies as "renormalization" calculations, which work by "finding one negative infinity for each positive infinity, so that in the sum of all possible contributions the infinities cancel" (p. 119), and "ghost particles" which, though they do not exist, are added to make the calculations come out right in the end. Although negative and positive infinities can be manipulated mathematically, it is very hard to connect these formal operations with an intuitively plausible reality.

PART II

LITERARY STRATEGIES

Chapter 2

DRAWN TO THE WEB
The Quality of Rhetoric in Pirsig's *Zen and the Art of Motorcycle Maintenance*

> He who without the Muses' madness in his soul comes knocking at the door of poetry, thinking that art alone will make him fit to be called a poet, will find that he is found wanting and that the verse he writes in his sober senses is beaten hollow by the poetry of madmen.
>
> Plato, *Phaedrus*

Robert M. Pirsig's version of the field concept derives in part, as his title suggests, from the Zen concept of a fluid, dynamic reality that precedes and eludes verbal formulation. Yet it is also informed by the Western tradition that sees the Word as the ultimate reality. The concern with language that was one of the keynotes of the last chapter is central to Pirsig's attempt to find a rhetoric capable of meeting these conflicting premises.

The emphasis on rhetoric is apparent in the "Author's Note" that introduces the narrative. In it, Pirsig claims that "what follows is based on actual occurrences," but adds that "much has been changed for rhetorical purposes."[1] In this ambitious autobiography that is also a novel,[2] three distinct rhetorical strategies are evident: those of the au-

[1]Robert M. Pirsig, *Zen and the Art of Motorcycle Maintenance* (New York: Bantam, 1980), p. 1. I use the Bantam page numbers since this is the most widely read edition. They can be converted to the page numbers in the Morrow edition (New York, 1974) by multiplying by 13/12.

[2]Apparently there are extensive parallels between the author's life and the biography he presents in *Zen*. He did teach at Montana, was a technical writer, and had a son who was institutionalized for a time for mental illness. This information on Pirsig's life is not firsthand; it comes from mutual acquaintances.

thor; those of the unnamed narrator, whose ideas obviously overlap with the author's, but who is also treated ironically; and those of Phaedrus, the shadowy other self that the narrator used to be. The three have in common the desire to find a rhetorical mode that will allow them to represent in words the field view of reality that they call "Quality." Contrary to what the "Author's Note" implies, rhetoric is not peripheral to this enterprise; it is at its center.

Phaedrus's approach to defining and disseminating this field view is as bold as it is naive. According to the narrator, Phaedrus was technically a genius, scoring 185 on the Stanford-Binet I.Q. test. His ambition, and his failure, were proportionate to his intelligence. His attempts to reform the entire structure of classical reason ended in a mental breakdown, a court-ordered institutionalization, and an eventual eradication of his personality by electroshock therapy. "He was dead," our narrator affirms.

> Destroyed by order of the court, enforced by the transmission of high-voltage alternating current through the lobes of his brain. Approximately 800 mills of amperage at durations of 0.5 to 1.5 seconds had been applied on twenty-eight consecutive occasions, in a process known technologically as 'Annihilation ECS.' A whole personality had been liquidated without a trace in a technologically faultless act that has defined our relationship ever since. I have never met him. Never will. (p. 77)

But Phaedrus has left behind a legacy—trunks of notes, recollections of him by family and friends, even fleeting memories that, like flashes of lightning, illuminate the narrator's quest for him. From these the narrator reconstructs Phaedrus's story; it centers on trying to understand the relationship between language and the holistic, dynamic reality that he calls "Quality."

Almost from the moment that Phaedrus conceives of Quality, he senses that it cannot be defined. His initial insight is confirmed when he has a sudden intuition that what he has been calling Quality is the same as the Tao of classical Zen thought. As he reads through his handwritten copy of the *Tao Te Ching,* he makes a "certain substitution" that confirms his insight: "The quality that can be defined is not the Absolute Quality" and "The names that can be given it are not Absolute names" (p. 227).

But Phaedrus, teaching rhetoric at the University of Montana, is pressed by academic colleagues for a definition. Under pressure as much from his own commitment to reason as from his fellow English teachers, he decides to risk a definition, proclaiming that Quality is the moment when subject and object meet, the instant of "preintellectual awareness" from which flow all of our conscious images of the world. The reader will recognize in this formulation a model very similar to the one Bohr proposed in his interpretation of the Uncertainty Principle. Pirsig, however, chooses to locate Phaedrus's response as part of the much earlier tradition of Western rationalism. "Why he chose . . . to respond to this dilemma logically and dialectically rather than take the easy escape of mysticism, I don't know," the narrator confesses.

> But I can guess. I think first of all that he felt the whole Church of Reason [Phaedrus's term for academe] was irreversibly *in* the arena of logic, that when one put oneself outside logical disputation, one put oneself outside any academic consideration whatsoever. Philosophical mysticism, the idea that truth is undefinable and can be apprehended only by nonrational means, has been with us since the beginning of history. It's the basis of Zen practice. But it's not an academic subject. (p. 207)

The decision marks a turning point. From there Phaedrus's path takes him to the University of Chicago to write a doctoral dissertation on Quality. At Chicago he enrolls in "Ideas and Methods 251," a course in classical Greek rhetoric. Already tending toward megalomania and paranoia, Phaedrus sees in the Chairman's conduct of the class a plot to defeat the rhetoric whose champion Phaedrus conceives himself to be. The plot is appropriate, for in pitting the Aristotelian Chairman against him, it re-enacts the struggle Phaedrus imagines took place in ancient Greece between the rhetoricians and dialecticians, which in his view was a struggle over whether reality could or could not be captured in words.

The narrator presents Phaedrus's reconstruction of Greek thought at face value, but this highly conjectural scenario is of interest more for what it reveals about Pirsig's text than for what it teaches about Greek history.[3] According to Phaedrus, the Sophists, dedicated to rhetoric,

[3]Phaedrus's reconstruction of Greek thought comes in for some hard knocks from an anonymous reviewer in the *Times Literary Supplement,* who intimates that the narrator's more egregious errors (for example, defining "Phaedrus" as "wolf") are owing to the American habit of reading the classics at third-hand remove ("On the Road with Aristotle," *Times Literary Supplement* No. 3763 [April 19, 1974], 405–406). American reviewers, on the other hand, tend to attribute these errors to the misapprehensions of the self-taught; see, for example, George Steiner's fine review, "Uneasy Rider," *The New Yorker* (April 15, 1974), 149–150.

had already formulated an idea of Quality, which they called "the Good." Like Quality, the Good "was not a *form* of reality. It was reality itself, every-changing, ultimately unknowable in any kind of fixed, rigid way" (p. 342). Because it cannot be known directly, it must be presented through analogy. The purpose of rhetoric is to create the analogies that can awake the apprehension of the Good in the listener's mind. To Plato and the dialecticians, however, reality was not the dynamic interaction the rhetoricians believed it to be, but a "fixed and eternal and unmoving Idea" (p. 342). Hence it can be spoken directly, without need for analogy; the proper tool for its representation is not rhetoric but dialectical analysis.

From this initial schism between the Good and the True evolve the modern dichotomies that are the subject of the narrator's discourse. When the Truth-lovers won over the Sophists, the narrator conjectures, Western civilization was started on the path that led to stunning technological feats, but emotional and aesthetic sterility. In this long decline into a society that believes in doing what is reasonable even when it isn't good, rhetoric is demoted from that which is best suited to represent the Good, as the Sophists see it, to the illegitimate emotional persuasion that Plato alleges it to be, and finally to the classification to which Aristotle consigns it, a branch of pandering.

This long, pseudo-philosophical disquisition has a suspense not easily conveyed here, for running alongside Phaedrus's reconstruction of Greek thought is his own battle with the Chairman.[4] After some preliminary skirmishes, Phaedrus finally defeats the Chairman on a point which any rhetorician instinctively appreciates, but which Truth-lovers tend to overlook: that the spoken word is only an analogy to reality, not reality itself. Seated at a classroom table that has a crack running down the middle, in keeping with the cultural schism being re-enacted there, Phaedrus defeats the Chairman by locating in the Platonic dialogue from which his name is taken the moment when Socrates admits that

[4]This scenario suggests that the author is innocent of knowledge about developments in the philosophy of science since the late 1800s. This is not the only example of such naïveté; whenever the narrator attempts a discussion of the history of philosophy, he betrays what George Steiner calls "potted summaries" of very complex issues (Steiner, p. 149). That he should nevertheless be concerned with issues that have dominated the philosophy of science in this century is striking evidence that the cultural matrix is capable of guiding individual inquiry in parallel directions, even when there is little or no direct influence between the different inquiries.

his parable of the chariot drawn by two horses is not truth itself, but a representation of truth. Thereafter Phaedrus regards the Chairman with a mixture of contempt and pity; in his mind, his triumph has reversed the ancient triumph of dialectic over rhetoric.

But though Phaedrus believes he won the battle, he finally comes to see that he has lost the war, for he "is doing the same bad things himself" as the dialecticians do when they use words as if they were reality.

> His original goal was to keep Quality undefined, but in the process of battling against the dialecticians he has made statements, and each statement has been a brick in a wall of definition he himself has been building around Quality. Any attempt to develop an organized reason around an undefined quality defeats its own purpose. The organization of the reason itself defeats the quality. Everything he has been doing has been a fool's mission to begin with. (p. 357)

Thereafter he turns to silence, sitting in the corner of his bedroom letting his urine flow naturally, letting his cigarette burn down naturally until it is extinguished by the blisters forming on his hand. Depending on one's viewpoint, this state can be seen either as a mystical ecstasy in which Phaedrus is finally at one with the Quality moment, or as a withdrawal into the insanity that the narrator so much fears. Perhaps the two are indistinguishable.

Though Phaedrus's failure is an extremely poignant moment, on reflection we can see that failure was the only possible outcome of his struggle with the University. That this realization is apt to strike us only after we have finished reading testifies to the narrator's evocative skill. But to try to imagine Phaedrus actually writing his dissertation on Quality is to realize how futile the effort must have been. The proposition that Quality could be defined in a dissertation, let alone defended, is apt to inspire incredulity in anyone who has experience with dissertations. Phaedrus fails because he cannot find a suitable rhetorical mode in which to embody Quality. Committed to reason, he cannot resist being drawn into definitions and dialectical argument, and he then inevitably loses the Quality he pursues.

The failure is not, however, the end of the quest to capture Quality in words. Pirsig's narration is a fresh start from a different direction. Pirsig, cannier and more wary, begins with the recognition that analytical discourse alone is not enough; his narrative differs from Phaedrus's aborted dissertation in its fuller use of rhetorical resources. In a sense, the dissertation has been written after all; but it is now combined with the emotions that electrify Phaedrus's quest, and encapsulated within the philosophical discourse that Pirsig calls his "Chautauquas." Surrounding the narrator's intellectual, abstract analysis of Quality is an extraordinarily complex rhetorical superstructure—all the more complex because it poses as a simple transcription of events.

Like Phaedrus, the narrator's focus is on reason. "About the Buddha that exists independently of any analytical thought," Pirsig writes, "much has been said—some would say *too* much, and would question any attempt to add to it. But about the Buddha that exists *within* analytic thought, and *gives that analytic thought its direction,* virtually nothing has been said" (p. 70). The goal, then, is not to abandon rational thought, not to attempt, as the Zen *koan* does, to involve the conscious mind in contradiction and paradox until it gives up and comes to rest. Rather, the attempt is to combine rational analysis with a fuller use of rhetoric so that the reader experiences Quality even while hearing about it. The means by which the narrator attempts this synthesis is deceptively simple: an alternation between past- and present-tense narration. The narrator begins, for example, by saying "I can see by my watch, without taking my hand from the left grip of the cycle, that it is eight-thirty in the morning," but then moves into the past-tense narration characteristic of the Chautauquas. The narrative thus proceeds in two different modes: the narrator's evocative descriptions of the immediate scene, and the analytical discourse of the Chautauquas. The divisions correspond with what the narrator identifies as the Romantic and Classic modes of understanding, one based on an intuitive appreciation of immediate surface, the other on an intellectual analysis of underlying form. At the very beginning of his tale, the narrator remarks that he prefers motorcycles to cars because "on a cycle the frame is gone. You're completely in contact with it all. You're *in* the scene, not just watching it any more, and the sense of presence is overwhelming" (p. 4). Talking about Quality in the Chautauquas helps us to understand the concept intellectually, while coming back to the "scene" maintains our ongoing relationship with the Quality moment.

Of course, this involvement is a rhetorical illusion. What the narrator tries to occlude from our immediate consciousness is the obvious fact that such descriptions are not experiences at all, but verbal reconstructions of sensory stimuli which may or may not have happened in the first place. The narrator's description of the "immediate" moment in fact embodies the very duality that the Quality event is meant to circumvent. As Pirsig defines the Quality moment, it is an undifferentiated unity that precedes and eludes intellectual concepts; it is therefore analogous to the turning kaleidoscope that we imagined in Chapter 1, whose fluid, inclusive dynamics defy classification into "patterns." But when the narrator writes as a person describing a world "out there," he has already bifurcated that fluid, dynamic whole into a subject regarding an object. What the narrator knows but does not admit is that even his immediate "scene" is an artifact that comes after the moment, a division imposed by the conscious mind as it seeks to understand the world as distinct from itself. As the moment that precedes intellectual awareness, the Quality event has passed by even before we read the narrator's present-tense descriptions. Between any verbal construct and the Quality event is a difference that is by its nature not sayable, because to speak inevitably implies that one is not the Quality moment but separate from it. At best language can only describe what has been, not what is.

That the narrator's rhetoric, though more complex than Phaedrus's, is still not adequate to the enormous task he sets himself becomes apparent as he keeps getting caught in the fundamental dilemma involved whenever Quality enters the realm of discourse. In the following passage, the narrator tries, as Phaedrus did with his students, to convince us that we already know what Quality is. Using his favorite metaphor of the mechanic who cares about and is involved in his work, the narrator describes the Quality experience.

> What produces this involvement is, at the cutting edge of consciousness, an absence of any sense of separateness of subject and object. 'Being with it,' 'being a natural,' 'taking hold'—these are a lot of idiomatic expressions for what I mean by this subject-object duality, because what I mean is so well understood as folklore, common sense, the everyday understanding of the shop. (p. 266)

The paradox of speaking Quality is implicit in the images the narrator uses to describe it. He talks about an "absence of any sense of separateness," but then identifies this awareness as taking place at the "cutting edge of consciousness." The knife imagery, as we shall see, occurs elsewhere as a metaphor for Aristotelian analysis. But the narrator too wields a knife when he speaks, as the "cutting edge" of his consciousness divides his pre-intellectual awareness of the event into the verbal abstractions of language. A variation of this dilemma appears in the narrator's repeated assurances that the ordinary people who are his readers already know what Quality is from "folklore, common sense, the everyday understanding of the shop." If he can achieve consensus, he can avoid defining Quality. But in order to achieve it, he must speak; his voice is what invites (or if we are more skeptical, creates) consensus by revealing to us how his thought and ours are the same. Consensus can be established, then, only by speaking; but speaking distorts the essence of the Quality that we are presumed to share. As the voice continues to enlarge the area of discourse, bringing more and more of Pirsig's thought into the common consciousness of reader and narrator, the problem only becomes more acute. For as the voice continues, more and more "bricks"—words, definitions, statements—stand between us and the Quality moment.

The narrator's problem with rhetoric is endemic to his narrative. The narrator warns that in classical Aristotelian analysis there is an "invisible knife moving," cutting the world into parts. But as we have seen, his own discourse does exactly the same thing, as his bifurcated narrative form suggests. Though this double form is an attempt to combine into one text both immediate experience and rational analysis, its effect is to further cut up into pieces the unity that Quality presupposes.

But the pursuit of Quality is only one goal of the narrator's speech. More pressing, and for him equally as important, is the need to prove his sanity. This he does by asserting his difference from Phaedrus. The narrator's pretense that Phaedrus is a person separate from himself is part of an elaborate defense mechanism, for we gradually realize that the narrator is the personality that emerged after Phaedrus's personality was annihilated by electroshock therapy. The narrator's relationship to this former self is intensely ambivalent. On the one hand he admires Phaedrus, spending countless hours attempting to reconstruct his ideas and planning a motorcycle trip so he can revisit Phaedrus's former haunts. But he also fears and flees from him, or more precisely from the possibility that this part of the self will return to assert that Pirsig, not Phaedrus, is the ghost.

For the narrator, the self has thus been artificially divided into a speaking subject and a passive object. If form is itself a message, then the message conveyed by this split narrative, and split narrator, is the same: his rhetoric is not overcoming duality, but reinforcing it. As he says when he discovers that the hairline fracture in his friendship with John is representative of a much larger schism within the culture, "You follow these little discrepancies long enough and they sometimes open into huge revelations" (p. 47).

I should like now to enlarge the framework of the discussion by referring to a distinction that the narrator rightly insists is crucial. "The application of this knife, the division of the world into parts," the narrator points out, ". . . is something somebody does. From all this awareness we must select, and what we select and call consciousness is never the same as the awareness because the process of selection mutates it. We take a handful of sand from the endless landscape of awareness around us and call that handful of sand the world" (p. 69). What Phaedrus and Pirsig seek is to "direct attention to the endless landscape from which the sand is taken" (p. 70). In a passage whose italics indicate his depth of feeling on the issue, the narrator insists that "it is necessary to see that *part* of the landscape, *inseparable* from it, which *must* be understood, is a figure in the middle of it, sorting sand into piles. To see the landscape without seeing this figure is not to see the landscape at all" (p. 70). The figure in our landscape, however, the figure we must see if we are "to see the landscape at all," is not the narrator but the author.

The narrator explicitly denies that his rhetorical intent extends beyond the simple strategy of a bifurcated narrative. "I suppose if I were a novelist rather than a Chautauqua orator," he writes, "I'd try to 'develop the characters' of John and Sylvia and Chris with action-packed scenes that would also reveal 'inner meanings' of Zen and maybe Art and maybe even Motorcycle Maintenance. That would be quite a novel, but for some reason I don't feel quite up to it" (p. 120). If the narrator is not quite up to it, however, the author is. As we shall see, increasingly Pirsig *is* developed as a "character" who engages in "action-packed scenes" that reveal a great many "inner meanings." Only when we turn to consider the author's rhetorical strategies do the full complexities of the attempt to capture Quality in words become apparent.

The subtlety of the author's rhetorical strategies can be seen in the ironies that emerge at the narrator's expense. The most important, perhaps, occurs in the narrator's relation to his son Chris. Richard H. Rodino notes that the narrator at the very beginning of the book "makes an *a priori* acceptance of the limitations of motorcycle travel that becomes a staggering threat to the Quality of his everyday life" when he commits himself to an internal monologue rather than an active interaction with his son.[5] "Unless you're fond of hollering you don't make great conversations on a running cycle" (p. 6). Lost in what Rodino calls "the cottony silence of his own thoughts,"[6] the narrator turns more and more from the real child on the back of his cycle to the hypothetical and abstract audience of the Chautauquas. It is from his readers that he hopes to gain the consensus that will validate his sanity and justify the Quality of his discourse over what he characterizes as the dangerously insane tirades of Phaedrus. But the deteriorating quality of his relationship with Chris shows that such introspective discourse works against Quality in his immediate surroundings, and ultimately against the consensus with his readers that he strives so hard to achieve. For as the ironies multiply, doubts grow in the reader's mind that Phaedrus was quite as inadequate as Pirsig claims, or that Pirsig is as fully adequate as he would have us believe.

In retrospect we can appreciate that the signals begin very early, for example in the narrator's remark, when he first begins to talk about Phaedrus, that "the purpose of the enlargement is not to argue for him, certainly not to praise him. The purpose is to bury him—forever" (p. 60). The narrator is not aware, at least consciously, of the Shakespearean echo; it is a signal not from the narrator, but from the author. How appropriate the irony is becomes apparent as the narrator continues his Chautauquas, for it is increasingly evident that the effect of his talking about Phaedrus is precisely the opposite of what he intends. Rather than "burying" Phaedrus, the narrator's discourse is resurrecting him, in more than one sense. From the viewpoint of the text as a

[5]Richard H. Rodino, "Irony and Earnestness in Robert Pirsig's *Zen and the Art of Motorcycle Maintenance*," *Critique*, 22 (1980), 24.
[6]Ibid.

rhetorical structure, for the narrator to talk about Phaedrus is to create him as a character; the only claim the narrator has to being more "real" than the Phaedrus he describes is that he is able to frame and encapsulate Phaedrus within his discourse. The first intimations that this strategy of encapsulation will not be successful come in Chris's repeated references to mysterious conversations that he has had with his father, but which Pirsig is unable to recall. Gradually it becomes clear to us—and eventually to Pirsig—that Phaedrus is breaking out of the frame of the narrator's discourse. The voice talking with Chris while Pirsig "sleeps" is not Pirsig at all, but Phaedrus.

The author's rhetorical strategy is perhaps now apparent. He has created a narrator who claims to be able to represent Quality within his discourse. At the same time, he has subtly involved the narrator in the contradictions that speaking Quality implies. But the narrator is only half of the persona; behind him, hidden from view and almost, but not quite, barred from discourse, is the shadowy Phaedrus. He is the part of the narrator, and the part of the narrative, that cannot be spoken. His nonetheless very real presence in the narrative haunts and animates it, as the Quality that eludes verbal formulation haunts and animates it. Phaedrus is the rhetorical analogue to the Quality that cannot be spoken.

We are now in a position to consider the narrator as what he insists he isn't: a character in a highly wrought, and at least partly fictional, rhetorical structure. The author's rhetorical strategy puts the ideas presented in the Chautauquas into ironic tension with complex image patterns that contradict, rather than extend, the intended message of the narrator's discourse. The effect of this tension is to draw into the discourse the central fact that the narrator tries to suppress: Phaedrus's existence. Two highly charged moments illustrate the technique. One is significant because in it a vital confrontation is avoided; it occurs when the narrator refuses to continue to the top of the mountain. The second occurs when Pirsig's quest ends at the ocean. Together the examples posit a central question. If the rhetorical strategies of Phaedrus and Pirsig are both revealed as inadequate to express Quality, how adequate is the author's strategy?

All along the narrator has talked in his Chautauqua about the "high country of the mind" and the "mountains of thought" that Phaedrus attempted to scale in his quest for Quality. His reaction to this awesome height is ambivalent; he clearly appreciates the grandeur of the mountains, but is himself more comfortable on the plains. When he commits himself to climb the mountain alone with his son, he is entering Phaedrus's terrain in both a metaphysical and a literal sense, for Phaedrus used to retreat to these mountains to help him crystallize the "mountains of thought" that Pirsig also attempts to scale. Moreover, now that John and Sylvia Sutherland (with whom they have been traveling) have left and there is no one around but the narrator and Chris, the tensions between them become more apparent. It is in this setting that Pirsig realizes the conversations Chris keeps mentioning are not Chris's childish fantasies or Pirsig's own incoherent mumblings, but the voice of the emerging Phaedrus.

The narrator's reaction to entering this emotionally charged terrain is complex. He knows that to climb the mountain is to invite a confrontation with Phaedrus, a prospect that he finds terrifying as well as potentially liberating. His response to this dilemma is to repress the conscious recognition that he is in some sense climbing to meet Phaedrus. So it is indirectly, through his Chautauqua during the climb, that we see the complexity of his reaction.[7] The Chautauqua is a discourse on "selfless" as opposed to "ego" climbing; Pirsig uses as his example Phaedrus's attempted pilgrimage to Mount Kailas in India. Though Phaedrus was physically stronger than those who came to the mountain to worship, he never made it to the top, while they did. Phaedrus, an ego climber, was trying "to broaden *his* experience, to gain understanding for *himself*" (p. 189). But for the selfless climbers, "each footstep was an act of devotion." The goal for them was not to reach the top but to participate in a process that reached its natural culmination at the mountain's peak. The narrator implicitly identifies himself as a "selfless climber" by comparing his attitude as he climbs to Chris's egoism. Chris had been to a summer camp where the emphasis was on achievement, and he climbs the mountain to prove how tough he is; reaching the top, not enjoying the climb, is his goal. Pirsig, on the other hand, concentrates on the present reality of each step, refusing to think beyond to the next. To give Chris an object lesson in selfless climbing, he allows the child to

[7]Rodino points out that many readers have been insufficiently sensitive to the author's ironic treatment of his narrator; he suggests that the obvious sincerity and earnestness of the narrator, as well as the book's own claim that it is "in its essence" fact, contribute to the problem. Readers, according to Rodino, are "reluctant to admit there might be anything artful or fictional" in this text (p. 21).

overextend himself so that Chris must either admit defeat or drive himself to exhaustion.

The narrator's attitude is apt to infuriate his readers, for what the reader sees is a father who deliberately drives his son to tears and rage, and then refuses to comfort him. But Pirsig's motives are more complex than this reading admits. Refusing to think beyond the present step is indeed what lets the narrator continue, for to anticipate the top would be to realize that each step takes him closer to confrontation with Phaedrus. So he goes along, one step at a time, always moving closer to the goal that his conscious mind cannot admit. His personality, terrified by the knowledge that it is not the complete self, would discontinue the climb if this goal were fully conscious; but the deeper self, desiring to heal the division within, keeps this knowledge from the narrator's consciousness. The contrast between ego climbing and selfless climbing is thus true in a sense that the narrator does not fully realize.

These conflicting desires are brought to the surface when Pirsig learns from Chris that Phaedrus's voice has told Chris he will be waiting for him at the top of the mountain. With the hidden goal now made explicit, Pirsig refuses to continue to climb. The psychological complexity of the refusal is enriched and brought into focus by the reason Pirsig gives for going back; he tells Chris he has "bad feelings" about spring rockslides. "Underneath us, beneath us right now," he says, "there are forces that can tear this whole mountain apart" (p. 218). As we shall see, resonating behind this remark are extensive metaphoric patterns of substance and motion that reveal the fallacies and contradictions in the narrator's stance.

The narrator likes to think that the metaphors of substance he appropriates to himself testify to his solidity, while the watery, insubstantial metaphors with which he surrounds Phaedrus confirm Phaedrus's nonexistence. But substance can be set in motion; and motion overcoming inertial mass is how the narrator describes Phaedrus's union with the Quality moment. A recurring nightmare for Pirsig is the fear that his substance will be buried, or carried away, by the same violent motion that swept Phaedrus into the "no-man's land" of insanity. When he visits Phaedrus's old office at the University of Montana, for example, he experiences an "avalanche of memory" (p. 160). As he advances farther into the room, he likens the returning memories to violent motion: "Now it comes down!" (p. 160).

For the narrator, this wild, uncontrolled motion has a double connotation that reveals the essence of his dilemma: it is associated *both* with Phaedrus's insanity and with the quest for Quality. For example, when Phaedrus has his mystical intuition that Quality and the Tao are one, the narrator describes the realization as if it were an avalanche.

> Then his mind's eye looked up and caught his own image . . . but now the slippage that Phaedrus had felt earlier . . . suddenly gathered momentum. . . . Before he could stop it, the sudden accumulated mass of awareness began to grow and grow into an avalanche of thought and awareness out of control; with each additional growth of the downward tearing mass loosening hundreds of times its volume, and then that mass uprooting hundreds of times its volume more . . . until there was nothing left to stand.
>
> No more anything.
>
> It all gave way from under him. (p. 228).

According to the narrator, then, Phaedrus's insanity began with a bifurcation of the self—the mind's eye detached from and observing "his own image"—and progressed like a rockslide to sweep him out of the mythos of his culture, into the no-man's land that society calls "insanity." When Pirsig refuses to continue up the mountain, what he fears is not the physical rockslide, but this mental avalanche.

But in other contexts substance in motion has a positive value for the narrator. One of the major faults he finds with classical Aristotelian analysis is that it cannot account for motion in the material objects it dissects. According to the narrator, the omission is crucial because, by preventing us from realizing the essentially dynamic nature of reality (in the field view), it consigns us to a dualistic universe in which motion and matter, mind and body, are separate and distinct. Determined to avoid this split, the narrator always chooses *moving* objects as his metaphors for Quality: the motorcycle in action, or the moving train of consciousness being guided by the track of Quality. Without this motion, the narrator asserts, the train is "static and purposeless": "A train really isn't a train if it can't go anywhere. In the process of examining the train and subdividing it into parts we've inadvertently stopped it, so that it isn't really a train we are examining. That's why we get stuck" (p. 254).

What the narrator fails to see is that in his anxiety to portray himself as a man of substance, he is separating himself from the metaphoric motion that, in other contexts, he recognizes as essential to Quality. The motion characteristic of Phaedrus's quest, like Phaedrus himself, has been consigned to a realm the narrator wants nothing to do with: waves of crystallization, avalanches of awareness, rockslides of memory. The result could be predicted, since it is the same fate that classical analysis suffers when it regards every material entity as static. The incipient division of self that Pirsig describes in Phaedrus has not disappeared. Rather, it has deepened, and two entirely different personalities have crystallized. The cautious part of the mind that retains its footing and observes the rest has become Pirsig; the part in violent motion, detached from society and consensus reality, is Phaedrus. The narrator, by identifying only with the substance and regarding motion as an alien quality, has not overcome the subject-object split. Rather, he has rendered it even more powerful by incorporating it into the structure of his personality.

The metaphoric patterns that help bring these psychological subtleties into focus reveal how much more sophisticated are the author's rhetorical strategies than those of the narrator. In the author's technique, form and content collaborate in a way they do not in the narrator's discourse, giving extraordinary depth and complexity to what one critic has called the narrator's "flat Midwestern" tones.[8] If Phaedrus is too abstract and esoteric, the narrator is too prosaic. It is neither one alone, but the two together, that infuse the narrative with Quality. The author's rhetoric, by revealing the inconsistencies in the narrator's attempt to speak Quality, brings the narrative as a whole closer to Quality by establishing the connections between Phaedrus and the narrator that the narrator himself would deny.

As these metaphoric patterns of connection become more concentrated, the narrator begins to accept that the confrontation with Phaedrus cannot be postponed indefinitely. In a way he almost welcomes it; his trip is a quest for Phaedrus as well as a flight from him. The ambivalence the narrator feels toward his alter ego becomes increasingly clear as he nears the ocean. Though he remembers from his dream that the Phaedrus-voice has told Chris he will meet him at the bottom

[8]That it is Phaedrus's presence that rescues the book from dullness has been observed by almost everyone who has written on this book.

of the ocean, he does not run from this encounter as he did from the mountain top; rather, he embraces it. "It's hot now, a West Coast sticky hotness . . . and I'd like to get to the ocean where it's cool as soon as possible" (p. 313).

The narrator's longing for the ocean is significant, for he has consistently identified Phaedrus with water and moisture. When he recalls Goethe's "Erlkönig," for example, he describes the ghostly pursuit as taking place by the ocean, though in Goethe's poem the setting is inland, with no mention of water. As Thomas S. Steele points out, this appearance of water in the poem "is read in from the end of the novel,"[9] for it is at the ocean that the final encounter between the two halves of Pirsig's bifurcated self takes place. As he nears the ocean, the Midwestern Pirsig meditates on its significance. "Coastal people never really know what the ocean symbolizes to a landlocked inland people," he muses, "—what a great distant dream it is, present but unseen in the deepest levels of subconsciousness" (p. 364). It is no wonder that Pirsig associates the ocean with Phaedrus. Nor is it surprising that he manifests considerable ambivalence toward the ocean; though he is attracted by its promise of cool relief, he suggests that actually to arrive will be to experience disappointment. When the "conscious images are compared with the subconscious dream there is a sense of defeat at having come so far to be stopped by a mystery that can never be fathomed" (p. 364).

As the end point of the journey, the "source of it all" (p. 364), the ocean brings into focus the ambivalence Pirsig has felt all along about arriving somewhere as opposed to just traveling. "Sometimes it's a little better to travel than to arrive," he remarks early in the journey (p. 103). Countering this affection for "just traveling" is the narrator's predilection for putting things in their proper sequence. John Stark has noted that the narrator "seeks to arrange correctly sequences of causes and effects";[10] the most physically immediate example is the arrange-

[9]Thomas S. Steele, "*Zen and the Art . . . : The Identity of the Erlkönig,*" *Ariel,* 10 (1978), 84.

[10]John Stark, "*Zen and the Art of Motorcycle Maintenance,*" *Great Lakes Review: A Journal of Midwest Culture,* 3 (1977), 50. The contradiction between the narrator's reliance on causality, and his vision of a Quality moment that precedes and negates causal interactions, is reminiscent of the contradiction between the strict causality of Newtonian mechanics and its undermining by quantum mechanics. The correlation suggests how rooted the narrator still is in the Newtonian world view.

ment of his journey as a linear sequence of points across the continent.

In many instances the inclination toward sequence goes almost unnoticed, because it is appropriate and commonsensical. When the narrator warns about out-of-sequence assembly in repairing a motorcycle, he is merely describing a common problem that most novice mechanics encounter. In other contexts the predilection is more obvious, because less expected. Many of Pirsig's major "discoveries" consist of determining the proper sequence of events, as when he figures out that the mythos preceded the logos, or that those who embraced the Good were displaced by those who believed in Truth. Though he pays homage to the virtues of just traveling and the importance of "lateral drift," then, he reveals himself as very concerned to discover and reinforce the proper *linear* sequence.

Why should linearity be so attractive to the narrator, despite his disclaimers? We may conjecture that the attraction originates in the narrator's anxiety to construct a linear sequence between Phaedrus and himself. Phaedrus is the self who existed at an anterior point; Pirsig is the self who occupies the present point in time. When the linear journey begins to break down, it signals the narrator's resignation to the fact that the linear relationship he has constructed between himself and Phaedrus must also dissolve.

The mounting tensions in his relationship with Chris accelerate this dissolving linearity. It is Chris who forces the narrator to revalue what the narrator considers a nightmare of non-linearity, the memory of a deranged Phaedrus who is so disoriented that, unable to follow ordinary directions to find the bunk beds his wife sends him to buy, he wanders aimlessly through grey, dusty streets. Chris, reacting more to the deteriorating relationship with his father than to their seemingly purposeless journey, begins a "strange, unworldly rocking motion, a fetal self-enclosure" that seems to shut the narrator out and be "a return to somewhere that I don't know anything about . . . the bottom of the ocean" (p. 360). "Remember the time we went to look for beds?" Chris asks. To Pirsig's astonishment, Chris remembers it as "fun." With this comes the narrator's realization that Chris is crying not for Pirsig, but for the lost Phaedrus: "It's *him* he misses" (p. 361).

Though the breakdown of the linear journey is a source of panic to the narrator, in other contexts he has given a different value to this kind of "lateral drift." The state of mind in which one is completely baffled and stopped is the moment when, according to Zen discipline, the mind is ready to receive new insight. It is when the mind is freed from the stricture of linear thought that it can respond to the guidance of Quality. "If your mind is truly, profoundly stuck," the narrator affirms, "then you may be much better off than when it was loaded with ideas . . . stuckness shouldn't be avoided. It's the psychic predecessor of all real understanding" (pp. 256–257).

So important to the narrator is the principle of "lateral drift" that he repeatedly uses it to structure his narrative. He begins with some apparently "minor" fact or event, then slowly brings it into focus, at the same time exploring its interconnections with other phenomenon. The Sutherland's dripping faucet, John's dislike of the beer-can shim, a chance remark from an elderly lady about quality—these are the small, everday occurrences that lead to major new insights. In each case they seem peripheral, timid, unimportant; but revaluing them begins a train of discovery breathtaking in its scope.

Several times the narrator is on the brink of recognizing that his own linear sequences are keeping him from seeing something important, especially in his relationship with Chris. After explaining that the South Indian Monkey Trap works because the monkey cannot revalue his freedom over the rice, the narrator confesses, "I keep feeling that the facts I'm fishing for concerning Chris are right in front of me too, but that some value rigidity of my own keeps me from seeing it" (p. 282). Finally, at the edge of a cliff by the ocean, his plans in chaos, the linear sequentiality of the Chautauquas broken by the prospect of impending mental collapse, the journey westward stopped by the margin of the sea, the narrator allows the fact that he has all along been suppressing to come into the center of consciousness: "In all this Chautauqua talk there's been more than a touch of hypocrisy. Advice is given again and again to eliminate the subject-object duality, when the biggest duality of all, the duality between me and him, remains unfaced. A mind divided against itself" (p. 363).

With that the narrator is ready for Phaedrus to emerge from his shadows. As he stands on the cliff, he feels a "sense of inevitability about what is happening." "I'm being pushed toward something," he realizes, "and the objects in the corner of the eye and the objects in the center are all of equal intensity, all together in one" (pp. 399–400). As Phaedrus emerges from the periphery, for the first time the narrator can hear his voice, though he does not immediately claim it as his own: "We're in another dream. That's why my voice sounds so strange." A few moments later, however, he opens himself to the full realization that he and Phaedrus are one person: "That's what Phaedrus always said—*I* always said—" (p. 370). As the lines of communication open between Phaedrus and Pirsig, they open also between father and son.

In his repeated dreams of the glass door, the narrator had always assumed it to be a private symbol. But now Chris also mentions the glass door, and "a kind of slow electric shock" (p. 369) passes through the narrator, a faint echo, perhaps, of the electric shocks that annihilated Phaedrus.[11] Earlier the narrator, in a moment of depression, had wondered whether real communication was possible: "the idea that one person's mind is accessible to another's is just a conversational illusion, just a figure of speech, an assumption that makes some kind of exchange between basically alien creatures seem plausible" (p. 269). But now, when Chris identifies the door as the hospital glass through which Phaedrus last saw his family, the narrator realizes that it is not a solipsistic image, but a shared experience.

With that recognition Pirsig's memory joins that of Phaedrus. Where before there was a bifurcation between the two memories—Phaedrus's memory stopping at the glass door and Pirsig's extrapolated backward to the "party" he imagines he attended—the two now become one continuous whole: "It has all come together."[12] When the journey resumes, it is with a new sense of joy and purpose. For the first time father and son remove their helmets and talk together naturally, undoing Pirsig's original assumption that "you don't make great conversations on a running cycle" (p. 6). Amid these symbols of union and harmony, the narrator appears finally to have solved his rhetorical problems.

But has the author solved this? His rhetorical strategy has been to create a narrator who talks explicitly about Quality in an intellectual way but is simultaneously involved in situations that show he does not fully live Quality, however well he may understand it intellectually. The strategy allows the author to render dynamic the static intellectual dis-

[11]Thomas Steele argues convincingly that the ending reverses, point by point, Pirsig's earlier failures to communicate (Steele, pp. 90–91).

[12]John Stark uses the dislocation between the two memories to suggest that there are *two* bouts of mental illness, but the "party" memory is more likely a rationalization.

course of the Chautauqua, creating a series of strong internal tensions between what the narrator says and what he lives. As a result, the narrative becomes far more densely textured than is the discourse of either Phaedrus or Pirsig alone. The dichotomy within the narrator especially is a master stroke, for it allows the author to hint at the ineffable without having to speak it. The discourse thus operates on many levels at once: as intellectual inquiry; as a physical and spiritual quest; and as a dramatic embodiment of Quality as it were *between* the characters, in the unspoken tensions between Pirsig and Phaedrus.

Despite this inspired stratagem, however, the author has not escaped the central dilemma. As we have seen, the thrust of the narrative, from the first pages on, has been toward synthesis: synthesis between art and technology, between Classical and Romantic modes of understanding, between thought and feeling, and most important, between the speaking subject and passive object into which Pirsig has made himself and Phaedrus. When the narrator finally accomplishes the internal synthesis that makes him again a whole person and a responsive father, the design is carried to its logical conclusion. This is its triumph—and also its most significant limitation. The completion of the design has been accomplished by moving what had been peripheral into center consciousness, but at the cost of losing the periphery that had been the text's greatest strength. As Phaedrus joins with Pirsig, and as they speak again with one voice, there is nothing left unsaid, no aspect or part of Quality that has not been drawn into the realm of discourse. Hence the synthesis that allows formal closure also sabotages the text's rhetorical strategy of making the hidden Phaedrus the rhetorical analogue to the unspeakable Quality.

This failure accounts, I think, for the uncharacteristic murkiness at the end. The symbols have obviously been carefully chosen to indicate synthesis. The cliff recalls the mountain top Pirsig did not reach, while the ocean waiting at the foot provides the new element necessary to make the situation echo, not repeat, the earlier retreat from confrontation. Pirsig had earlier contrasted the "mountains of achievement" with the "ocean trenches of self-awareness" (p. 264), suggesting that both are necessary to make a culture or a life complete. On the cliff overhanging the sea, the two come together. As Pirsig and Chris arrive, the cliff is "surrounded by banks of fog," recalling the fog in Pirsig's retelling of the "Erlkönig." More subtly, the fog shrouds the scene in a kind of twilight, creating an ambiguous light that is halfway between the daytime when Pirsig rules and the night when Phaedrus speaks. After such a powerful concatenation of symbols, the author is almost obliged to suggest that the union is full and complete, the quest at an end. But Pirsig is too honest a writer not to acknowledge also that such quests for self-knowledge can never really reach a point that can be proclaimed "the end"; self-awareness is not a single goal, but a continuing process. So at the end the author tries to renege, implying that the goal has been reached and yet also suggesting that the journey is unfinished. As Pirsig and Chris continue on their trip, the narrator acknowledges that "trials never end, of course. Unhappiness and misfortune are bound to occur as long as people live" (p. 372); but at the same time he proclaims, "We've won it. It's going to be better now. You can sort of tell these things" (p. 373). The author can avoid having to deal with the paradox because here he ends his text. But strategic withdrawal at the point where the problems become insoluble, though certainly one of the options an author has (as we shall see in the next chapter with D. H. Lawrence and *The Rainbow*), does not solve the deeper underlying problem.

The narrator's desire for synthesis and completion is, I suspect, a less sophisticated version of the author's own drive toward closure. The mind at work here—whether that of author or narrator—clearly has a very strong bias toward order, synthesis, and union. That it should be fascinated by the possibilities of a field concept of reality is therefore not surprising. At its best, *Zen and the Art of Motorcycle Maintenance* gives powerful expression to the harmonies that the cosmic web can suggest: "Peace of mind produces right values, right values produce right thoughts. Right thoughts produce right actions and right actions produce work which will be a material reflection for others to see of the serenity at the center of it all . . . —a material reflection of a spiritual reality" (p. 267). But because of its lingering problems, *Zen and the Art of Motorcycle Maintenance* is important as much for the questions it raises as for the answers it posits. In devising a rhetorical strategy to cope with the paradoxes that arise when one attempts to speak from within the field, it has raised what is perhaps the most important issue for a literature that attempts to embody this view. That it finally yields to its own consuming desire for order means that, at the end, rational synthesis wins out over the ineffability of the whole.[13]

[13]For a very different valuation of Pirsig's inclination toward reason, see William

In this light, the epigram that Pirsig chooses for his text has perhaps unintentionally ironic overtones. It comes from the *Phaedrus:*

> And what is good, Phaedrus,
> And what is not good—
> Need we ask anyone to tell us these things?

By attempting to "tell us these things," Pirsig indeed may have described the Buddha that lies within rational thought—but at the expense of the Buddha that cannot be spoken. What Pirsig knows, but cannot fully accept, is that (as Heisenberg said of science) literature is not about reality but about what we can say about reality. In allowing the distinction to become blurred between his verbal representation of the field and the field itself, Pirsig in the end draws back from his encounter with the paradox at the heart of the cosmic web. For a full exploration of what it means to try to speak the ineffable, we shall have to wait until the final chapter, on Thomas Pynchon. In the meantime, we shall turn to other writers who respond to the dilemma of trying to represent reality through a field model by transforming or subverting the model itself.

Plancher's "The Trinity and the Motorcycle," *Theology Today,* 34 (1977), 248–256.

Chapter 3

EVASION
The Field of the Unconscious in D. H. Lawrence

> At this point it must be asked why the classical paradigm is so difficult to give up *in toto* . . . the confusion that has for so long been evidenced in discussions about quantum mechanics, and the intense emotions that such discussions can evoke, suggest that more is at stake than simply the comfort and success of an older paradigm.
>
> Evelyn Keller, "Cognitive Repression in Contemporary Physics"

About the time that logical positivism was approaching its heyday in science, D. H. Lawrence set forth a theory that he claims would form the basis for an entirely new kind of science.[1] The basic premise of Lawrence's "subjective science" was that it is possible to apprehend reality directly from a set of symmetrically arranged "centers" in the body, without mediation from the conscious mind. In this "science," statements are confirmed not by independent observation or replicate experiments, but by appealing to the intuition of others who will verify statements from their own unconscious centers. When the centers come together in an interactive bonding, they become, in Lawrence's terminology, "polarities."

Lawrence's theory is so obviously at odds with what was known even in his day that it can scarcely be taken seriously as "science" of any

[1]Lawrence's clearest explication of what he means by his new science is in the "Foreward" to *Fantasia of the Unconscious,* in *Psychoanalysis and the Unconscious and Fantasia of the Unconscious,* ed. Philip Reiff (New York: Viking, 1960), pp. 53–58.

kind.[2] Yet Lawrence was, in his way, wrestling with some of the same issues that were occupying the attention of contemporary science. His "subjective science" is an attempt to define a field of interaction that includes both subject and object. For Lawrence, the "field" is always identified with a breakthrough into what he calls the "unconscious." In order to reach the "unconscious," from which the "field" originates, the body centers of one person engage those of another in a fierce dialectic that ends when the two "polarities" come together in mystical union. Lawrence is evidently adopting his idiosyncratic terminology from the field theory he was most familiar with, Maxwell's theory of electromagnetic fields. As we saw in Chapter 2, Einstein credited Maxwell with beginning a transition in scientific thought which would lead, finally, to Einstein's claim that "the field is the only reality." Though Lawrence is very much following his own path, his attempts to define a psychological "field" clearly parallel these developments in science. Essentially ignorant of post-Newtonian physics, Lawrence nevertheless has a notion of an integrating field and understands that it must, by its nature, resist articulation. It is perhaps not surprising, then, that most physicists would agree (though for very different reasons) with many of Lawrence's deepest beliefs: that reality is a dynamic flux rather than the manifestation of rigid laws; that the observer, rather than being isolated in Cartesian objectivity, participates in that flux; and that certain aspects of reality will always elude deterministic analysis. Though ignorant of much factual knowledge about the new science, Lawrence anticipated the spirit of its principal results.

Beyond this parallelism lies a deeper connection between Lawrence and the new science, a connection illuminated by the strategies of resistance that Lawrence employed to oppose entry into what he ostensibly sought. At the heart of this connection is an intense ambivalence toward the concept of a field that unites subject and object and that hence tends to blur the boundaries between the self and other. In Lawrence, the ambivalence is so close to the surface that it can scarcely

[2]James C. Cowan's account of Lawrence's physiology demonstrates that it contradicts even what was known in his day about the workings of the nervous system. Neither the sympathetic nor volitional (autonomic) nervous system can apprehend directly; both send their messages to the appropriate cortical centers for processing. See particularly Cowan's chapter on "Lawrence's Romantic Values" in *D. H. Lawrence's American Journey: A Study in Literature and Myth* (Cleveland: Case Western Reserve University Press, 1970), pp. 15–24.

be missed; among quantum physicists it is less obvious, but no less invested with psychological complexities of the kind that make Lawrence's encounter with the field concept a potent force in shaping his art.

One indication that quantum physicists, like Lawrence, resist the field concept is their reaction to the Uncertainty Relation. In a provocative article, "Cognitive Repression in Contemporary Physics,"[3] Evelyn Keller notes that after fifty years of debate there is still no single accepted interpretation of what quantum theory implies about the nature of reality. After careful analysis she concludes that the so-called "Copenhagen Interpretation" of the Uncertainty Relation is an umbrella term "under which a host of different, often contradictory positions co-reside." We saw in Chapter 2 that Bohr and Heisenberg, though taking very different positions on the Uncertainty Relation, are nevertheless perceived by the scientific community as being of one mind on the matter. Even Heisenberg thought they were in agreement (though Bohr knew better). Keller argues that such extraordinary confusions and conflations provide "*de facto* evidence of defense and evasion."[4] What is being evaded, Keller suggests, is the recognition that the self exists neither in isolation from the world nor in mysterious sympathy with it. The vocabulary she uses to describe the struggle of physicists to come to terms with a reality in which "the boundaries between subject and object are . . . never quite rigid" is hauntingly familiar when applied to Lawrence: ". . . the capacity for objective thought and perception is not inborn, but rather . . . acquired as part of the long and painful struggle for psychic autonomy—a state never entirely free from ambiguity and tension. The internal pressure to delineate self from other . . . leaves us acutely vulnerable to anxiety about wishes or experiences which might threaten that delineation."[5] Elsewhere, Keller links the disinclination of scientists to admit to such an ambiguous reality to the process of gender differentiation, arguing that the male child, in our gendered culture, comes to see the mother as essentially different from himself. This gender differentiation then lays the foundation for the scientist's later objectification of the archetypal

[3]Evelyn Fox Keller, "Cognitive Repression in Contemporary Physics," *American Journal of Physics,* 47 (August 1979), 718–721.

[4]Ibid., p. 718.

[5]Ibid., p. 721.

female, Mother Nature.[6] When nature manifests herself as neither objective nor subjective but as a union of the two, therefore, the ambiguity is deeply troubling because it is not simply an intellectual issue, but an emotional crux connected to the deepest layers of the scientist's self-concept. Keller's argument thus supposes that there are fundamental and deep-seated connections between the process of gender differentiation in infancy, the subsequent strongly male orientation of scientists (including female scientists), and the resistance of scientists toward the field concept.

Lawrence's reaction to the integrated field of the "unconscious" is uncannily similar to the dynamic Keller imagines for quantum physicists. Like modern physicists, Lawrence is "acutely vulnerable" to anxiety when subject and object begin to merge; for Lawrence, the anxiety is most apparent when he imagines a son separating from his mother. This chapter will explore Lawrence's strategies of approach and avoidance toward the undifferentiated field of the "unconscious" and relate them to the process of gender differentiation that was, for Lawrence, the central issue of child development. Placing Lawrence in this context will illuminate his uneasy relation to the intellectual revolution of his time, and will allow us to prove some of the deeper reasons why even to today's physicists a field view can be threatening as well as liberating.

The question of how anxiety about the "unconscious" shapes Lawrence's art cannot be separated from what he called his "metaphysics," for Lawrence saw his creative writing and his polemical tracts as two sides of the same coin. Critics who address the relation between Lawrence's "metaphysic" and his art tend to fall into two camps: those who, like Frank Kermode, take metaphysic to be central;[7] and others, for example Leo Bersani and Colin Clarke, who see in the fiction a conflation of apparent contraries that wreaks havoc with Lawrence's metaphysical schematic.[8] To ask which position is correct is to ask the wrong question, for Lawrence's metaphysic is the cognitive version of a deeper paradox that emerges in a different way in the "confusions" of his fiction. The more fruitful line of inquiry is to ask what it is that is being

[6]"Gender and Science," *Psychoanalysis and Contemporary Thought,* 1 (1978), 409–433.

[7]Frank Kermode, *D. H. Lawrence* (New York: Viking Press, 1973).

[8]Leo Bersani, *A Future for Astyanax* (Boston and Toronto: Little, Brown, 1976), pp. 156–185; Colin Clarke, *River of Dissolution* (New York: Barnes and Noble, 1969), esp. pp. 49–69.

simultaneously revealed and concealed in both the art and "polyanalytics," and what these strategies of approach and avoidance can tell us about the underlying psychodynamics.

That the metaphysic, despite its polemical and revelatory stance, is concealing something is suggested by the instability of its dialectic. Lawrence repeatedly pays allegiance to the belief that reality is a dynamic whole and that we have the means for grasping its nature intuitively and directly. But the approach to this reality proceeds by a characteristic motion that is also a retreat from it. The breakthrough to the "mystic body of reality" is supposed to occur when two "polarities" are locked together in tense, dynamic interplay. With first one, then the other dominant, the "polarities" engage in a "frictional to-and-fro" that could, Lawrence believed, break through to an unbounded space that encompasses all opposites. This dialectic is extremely unstable, however, because one of the two "polarities" is consistently valued over the other. (Lawrence would of course argue that his privileging of the centers of resistance was merely in redress of society's emphasis on the centers of attraction.) Nevertheless, because of the differences in value, there is always an impetus to resolve the tension in favor of the more "dynamic" of the two terms. If the favored term prevails, the dialectic collapses into unity, leading, in Lawrence's terminology, to "rigidity." This incipient collapse can be prevented only if the two terms are subsumed into a larger unity which then becomes the favored term of a new dialectic. When this dialectic also threatens to collapse toward the favored term, it must be subsumed into another, still larger term. Only through successive enlargements can the dynamic be continued, and its continuation implies that its putative goal—the breakthrough into the unconscious—is never achieved.

Paradoxical as it is, this dissolving dialectic is only the first level of a deeper paradox that emerges when Lawrence struggles to move from this abstract scheme to its application in family relationships. According to the metaphysical scheme, the purpose of the dualistic to-and-fro is to reach the unconscious, that part of the psyche which is able to hold opposites in a continuing tension without needing to resolve them. But to enter this realm is also to encounter an experience that psychologists identify with the earliest stages of infantile consciousness: the lack of differentiation between self and other, specifically between the self and mother. The experience that in one sense is the desired culmination of

Lawrence's "frictional to-and-fro" is thus closely linked with a state he regards with horror, the child's fusion with the mother. On this deep level, the collapse of the "polarities" is surrounded by intense anxiety; in his psychoanalytic essays, Lawrence associates it with the "ghoul" of repressed incest desire.

Countering the positive connotations with which Lawrence surrounds the breakthrough into the unconscious, then, is an extremely strong anxiety about the loss of individuation this would entail. Read in this way, Lawrence's insistence in his novels that two "polarities" can fuse into each other and yet somehow still retain their individual autonomy is a strategy for introducing differentiation at precisely the point where it is under the most pressure to succumb to the undifferentiation of the unconscious. Such "confusions," far from being extraneous to the art, are the enabling strategies that allow it to go forward.

The structure of *The Rainbow* reveals how these "confusions" emerge and develop. Through the chronicle of the Brangwens, Lawrence attempts to depict modern man's fall into consciousness.[9] The ease with which the plot can be rendered as a schematic of increasing alienation shows how powerful the metaphysic is in organizing our experience of this text. Two patterns are apparent. Within each generation there is a dipolar interaction between the man and woman that is the "entry into another circle of existence."[10] When Tom and Lydia engage in this tense opposition of contraries, for example, they "open the doors, each to the other . . . it was the transfiguration, the glory, the admission" (*R*, p. 91). Countering this active dipolarity, however, is a linear decline through the generations. As child succeeds parent and as the society becomes more "conscious" and "mechanical," the partners are less and less able to engage each other in the "frictional to-and-fro" that is the key to the doorway of the unconscious.

This then is the formal pattern, the metaphysical schematic that supposedly dictates the arrangement of the material. In it we can see the instability that is characteristic of Lawrence's dialectic, as the "to-and-fro" motion is increasingly imperiled by the linear decline. But this instability is not all that interferes with the breakthrough to the "unconscious." Also present are many details that refute or heavily qualify the

[9]Leo Bersani (pp. 175–180) makes this point in discussing *Women in Love*.

[10]D. H. Lawrence, *The Rainbow* (New York: Penguin, 1976), p. 91. In further page references to this edition I will abbreviate it *R*.

notion that such a breakthrough is desirable in the first place. According to whether the reader attends to the schematic or the immediate texture, the text thus can appear as highly determined or inchoate. What we miss when we concentrate on either one alone is the way in which the interaction between the schematic and the "confusions" operates according to a symbolic logic of its own.

The logic is implicit in the to-and-fro which is a retreat from as well as an advance to the unconscious. This approach/avoidance is almost always bound up with a simultaneous identification with and rejection of the parent. As we shall see, such identification is not accidental, because the deeper struggle is between the drive to attain a fully individuated, autonomous state, and the nostalgic desire to fuse with another in a re-creation of the infant's identification with the mother. It is with the second generation, when we see the protagonists as both adults and children—that is, as both fully individuated beings and continuations of the parental consciousness—that the ambiguities surrounding the entry into the "unconscious" really begin to take hold. By the third generation, when the parental images are not one but two layers deep, the to-and-fro dialectic is so "confused" as to be unsustainable.

This argument has been partly anticipated by Colin Clarke, who has written persuasively on Lawrence's simultaneous aversion and attraction to what Clarke calls "reductive energy." Clarke notes that it is increasingly difficult for the reader to make distinctions that the narrator nevertheless insists are crucial: Will Brangwen's vulnerability which is also power; Ursula's "fierce salt-burning corrosiveness under the moon" which is at once freeing and destructive; and the "corrupt African potency" of Skrebensky which is both powerful and depraved. Distinctions which ought to be of "some thematic importance" are, Clarke argues, in the "final effect of the novel," played down.[11] Clarke interprets these "confusions" as Lawrence's first attempts to articulate a holistic reality which cannot be bifurcated into either-or categories. Clarke does not make the identification with the Uncertainty Relation, but it too, of course, also points toward the inadequacy of either-or formulations. If the uneasiness of quantum physicists with the Uncertainty Relation is rooted in early childhood experiences, as Keller sug-

[11]The quoted phrases are from *River of Dissolution,* p. 45.

gests, then the consistent identification Lawrence makes between entering the "unconscious" and regressing to an infantile state by implication illuminates the resistance in the scientific community to the Uncertainty Relation.

For Lawrence, when the parent-child duality begins to fuse into a single figure, the ambivalence becomes so intense as to make even an approach toward the "unconscious" untenable. The pattern is apparent in the cycle of births, matings, and deaths that constitute the chronicle of the Brangwens. For Tom and Lydia, who are cast as progenitors and who therefore convey the least sense of being both parents and children, the passage into the "openness" of the unconscious is the least ambiguous. With this couple the values of openness and enclosure are consistent and straightforward: to "open" oneself is to participate in the joyousness of the unconscious, while to be "closed" is to remain isolated within the sterile boundaries of ego consciousness.

By the time the second-generation couple, Will and Anna, mature from children to adults, the passage to the unconscious has become considerably more complex. Though Will passionately wants to "open" himself to Anna, this "openness" is threatening to both of them. Moreover, Anna's refusal of Will leads not to enclosure, but to an apparently different, more sinister openness. In rejecting Will, Anna leaves him a "prey to the open, with the unclean dogs of the darkness setting on to devour him" (*R,* p. 166). Openness is thus ambiguous, and so is enclosure. We are told that Will always remains aware of "some limit in himself, of something unformed in his very being . . . some folded centres of darkness which would never develop and unfold whilst he was alive in the body" (*R,* p. 207). The imagery points to a center of "darkness," usually a code word for the unconscious; but this is a "folded" center, a potential openness that nevertheless remains encapsulated. The metaphors image a paradoxical space that cannot be assimilated into the schematic, a space that is at once infolded and open, threatening and liberating, isolated but potentially dynamic.

The corollary to the convolutions of Will's interior space is the inward-turning of Anna's fecundity; "if her soul had found no utterance, her womb had" (*R,* p. 203). Even though Anna cannot go *through* the doorway with Will, through their union she *becomes* a "doorway and a threshold, she herself. Through her another soul was coming, to stand upon her as upon the threshold, looking out, shading its eyes for the direction to take" (*R*, p. 193). But this interior space, like Will's, is paradoxical; it is at once a negation and a fulfillment. Anna locates her identity in her ability to bear children, so for her the womb-space is a highly charged signifier, capable of conferring (and creating) identity. But the necessary and inevitable end of this process is the emptying of that space when the child is born, so that what begins as fulfillment ends in negation as mother and child break apart into separate beings. The contrary claims of autonomy and dependence are so fragile that they can be balanced only at the moment of equipoise, when Anna is *both* "a doorway and a threshold." Most of the time, the breakthrough into the "unconscious" has become so bound up with fears of dependence and of an inability to differentiate the self from the other that an unambiguous response is not possible.

By the third generation, when the layers of parental identity are not one but two layers thick, the anxieties become correspondingly intense. When the child of Anna's womb, Ursula, crosses the womb-threshold and begins the long process of establishing an identity independent of her parents, the ambiguities that had characterized her parents' relationship deepen for her into contradictions. Ursula attempts to escape the thickening layers of parental identity by rejecting her mother as a role model, insisting that she will not become a fecund mother in her turn. Determined to hold onto her autonomy, the most Ursula can offer or receive from Anton is "a sense of his or of her own maximum self, in contradistinction to the rest of life" (*R*, p. 301). Though she has not herself become the enclosing parental space, she still pays a price for her freedom, for the narrator tells us she cannot break out of the ego-space of the self, "wherein was something finite and sad, for the human soul at its maximum wants a sense of the infinite" (*R*, p. 301).

But *has* Ursula escaped from becoming the mother in her turn? She discovers that the female inheritance is not so easily transcended. Caught in the paradoxes of a freedom that is also an imprisonment, Ursula re-creates the creative/destructive womb-space in her relationship with Anton. Under the influence of the moon that Lawrence identifies in the *Fantasia* as the cosmic pole of female assertion and autonomy, Ursula drains from Skrebensky, in a fierce kiss, his "distinct male" core (*R*, p. 321). In another moonlit night by the sea, Ursula finally admits that Anton has no independent existence, and he, in a symbolic return to the archetypal enclosure, curls into the fetal crouch of the womb. His posture is one way of signifying what Ursula later realizes, that he is her "creation": he "had never become finally real . . . she had created him for the time being. But in the end he had failed and broken down" (*R,* p. 493).

The conflation of images thus moves into deeper contradiction as the schematic moves toward linearity. The more the end of this mechanical, repetitive society appears predetermined, the more the imagery insists on a merging of contraries that is anything but linear. As the schematic leaves behind the to-and-fro dialectic, the imagery takes it up, folding in upon itself in increasingly opaque convolutions. It is as if Lawrence were compelled to articulate a simultaneous approach to and avoidance of the "unconscious" so that if the metaphysical scheme does not allow for it, the imagery must.

What Lawrence is wrestling with in his use of imagery is part of a more general problem with language. We have seen how the language of *The Rainbow* bifurcates between an abstract schematic of linear decline, most apparent when one takes a bird's-eye view of the plot, and a highly stressed conflation of images that is most apparent at the level of textual detail. In the schematic, Lawrence follows the simple formula of imposing linearity on top of dipolarity; the decline across the generations is linear, while the dynamic within each generation is dipolar. But this implies that each successive generation, though it still has some dipolarity within the male–female relationship, is further along the linear scale, and it is left to the imagery, working as it were in defiance of the schematic, to keep open the dual potential of promise and threat inherent in the to-and-fro dialectic. Lawrence attempts to reinstitute dipolarity at the end by having Ursula hope for rebirth; but this attempt to mediate between two contraries—this time his own hope and the hopelessness appropriate to the linear schematic—has struck many readers as an arbitrary, if not desperate, solution.[12]

Such a solution was bound to be unsatisfactory, for the problem lies not just in the novel's structure, or in its use of symbol, but in the nature of language itself. On the one hand, Lawrence feels deeply that

[12]F. R. Leavis, *D. H. Lawrence, Novelist* (London: Chatto and Windus, 1955), pp. 142–143; see also Graham Hough, *The Dark Sun: A Study of D. H. Lawrence* (New York: Capricorn, 1959), pp. 71–72. For a full review of the controversy see Edward Engelberg, "Escape from the Circles of Experience: D. H. Lawrence's *The Rainbow* as a Modern *Bildungsroman*," *PMLA,* 78 (1960), 103–113.

reality is essentially mystical and unspeakable, to be experienced rather than understood rationally. On the other hand, he is committed to depicting this ineffable reality in words. The closer he comes to rendering the unconscious in language, the closer he paradoxically comes to destroying its realization, because language is necessarily conscious. In the "scientific" essays, the problem appears in its most acute form. There Lawrence commits himself to making what is already an ineffable mystery not only verbally explicit, but also systematic and rationally plausible. There is thus in the essays a strong contradiction between what Lawrence says and how he says it.[13] According to what Lawrence says, reality is best and most fully apprehended directly through the body's sensual centers, without mediation from the mind at all; but he makes this claim in the extremely abstract and objectified mode of a "scientific" discourse.

What in the essays exists as a disparity between form and content is still present in the novels, but in the more amorphous form of the fiction, it has not rigidified into anything quite so definite as a contradiction. It would be truer to say that in the fiction it exists as a paradox. It evolves from Lawrence's belief that real knowledge is always sensual and immediate rather than mental, and his simultaneous endeavor to make us apprehend this through the verbal abstractions and stylizations of art-speech. Simply put, the paradox is this: to know is not to be able to say, and to say is to move from the reality of unmediated knowledge into abstraction. It is a dilemma that Bohr also recognized as fundamental, for to speak is to enter into the "either-or" conceptualizations that quantum theory, like Lawrence, was trying to escape. Further complicating this already complex dilemma is the psychological substratum that links the inability to differentiate between subject and object with early childhood experiences, and consequently with the highly charged issues of identification with, and separation from, the mother. As we saw in Chapter 2 in connection with scientific models, the key to the complexity is language.

Lawrence's attitude toward language is implicit in the dynamics of linearity and dipolarity that are at work in *The Rainbow* and that continue to figure importantly in his plot construction in *Women in Love*.

[13]In "The Beginning and the End: D. H. Lawrence's *Psychoanalysis and Fantasia*," *Dalhousie Review*, 52 (1972), Evelyn Hinz discusses the relation between form and content.

In one sense language represents abstraction, the transformation of immediate experience into mental conception. Lawrence's suspicion of language is reflected in the verbal reticence of his characters. For example, after Birkin and Ursula, in the "Excurse" chapter of *Women in Love,* come in touch through their lovemaking with what Lawrence calls the "mystic body of reality," they are reluctant even to acknowledge their experience in words. "It was so magnificent," Lawrence writes, "such an inheritance of a universe of dark reality, that they were afraid to seem to remember. They hid away the remembrance and the knowledge."[14]

Throughout *Women in Love,* both Birkin and Ursula distrust words; that Birkin demonstrates on occasion an over-fondness for his own words is one of his weaknesses. Ursula is wiser in feeling "always frightened of words, because she knew that mere word-force could always make her believe what she did not believe" (*WL,* pp. 428–429). Birkin, despite his verbosity, shares her feeling. As Birkin is telling Ursula that they must go beyond the merely personal into some new sort of relation, Lawrence says that "she knew, as well as he knew, that words themselves do not convey meaning, that they are but a gesture we make, a dumb show like any other" (*WL,* p. 178). Lawrence's attempt to use language to move beyond language is apparent in the paradoxical imagery, in which language becomes a nonverbal "gesture" or even a "dumb show." When Birkin turns away "in confusion" because he cannot find the right words, Lawrence editorializes: "There was always confusion in speech. Yet it must be spoken. Whichever way one moved, if one were to move forwards, one must break through. And to know, to give utterance, was to break a way through the walls of the prison as an infant in labour strives through the walls of the womb" (*WL,* pp. 178–179).

The association of the womb with the need to give utterance is important, for it provides a key link between language and the differentiation of the self from the mother. The informing tension is between the desire to "move forward" and the fear that forward motion in language can be dangerous because it leads away from direct experience into mental experience and therefore to falsity. Lawrence's response to the dilemma of wanting to go forward and yet fearing the forward movement as a progressive abstraction is to imagine a movement of

[14]D. H. Lawrence, *Women in Love* (New York: Penguin, 1976) (henceforth *WL*), p. 313.

speech analogous to the contractions of labor, a rhythmic and tense pulsation capable of propelling one into a new existence. In this pattern the womb metaphor is central; it is an image to which we will return. For the moment, we can note that it is used here to suggest a dipolar rhythm that can counter the inherent linearity of language, thus permitting a forward motion without getting lost in abstraction.

By the time of the writing of *Women in Love,* Lawrence is articulating the dynamic explicitly. The "Foreword" sounds the keynote for the change. "In point of style," Lawrence writes, "fault is often found with the continual, slightly modified repetition. The only answer is that it is natural to the author; and that every natural crisis in emotion or passion or understanding comes from this pulsing, frictional to-and-fro which works up to culmination" (*WL,* p. viii). Through the "continual, slightly modified repetition," the linear flow of language is partially checked. The result is not straightforward linearity but the "frictional to-and-fro" of repetitive clauses that build through a series of periodic sentences up to the culmination of the single, short declarative sentence which—if the style works—is the point of breakthrough.

The style is an attempt, then, to make language somehow engage in a "frictional to-and-fro" that can break out of the envelope of ordinary perception to a direct apprehension of reality. Here the difference between Lawrence and the quantum physicists surfaces most clearly, for they would never admit such a mystical apprehension as a valid subject for their discipline. Bohr, for example, repeatedly emphasized that science is not about reality, but about what we can say about reality. For Lawrence, almost the opposite is true. For him literature is not what we can say about reality, but about what we cannot say about it; language is important only insofar as it can re-present the reality that lies beyond words.

To prefer words to reality is the mistake Gudrun and Loerke make in their "quips and jests and polyglot fancies." "The fancies were the reality to both of them," Lawrence writes scornfully; ". . . they were both so happy, tossing around the little coloured balls of verbal humour and whimsicality" (*WL,* p. 460). Lawrence is not interested in this kind of verbal intricacy because he fears that whatever calls attention to a particular verbal formulation can be dangerous, tempting the reader to stay on the verbal surface rather than go beyond the language to the reality to which it is meant to point. So Lawrence, having arrived at one way of saying something, keeps repeating it until the building tension explodes into what he hopes will be the reader's direct apprehension of the idea. At the same time, the "continual, slightly modified repetition" creates a movement of thought that is designed to minimize the inherent problem of using language by creating an internal tension between the back-and-forth prose rhythms and the syntactical linearity.

An excerpt from *The Rainbow* will illustrate how Lawrence anticipates the technique that was to find its full realization in *Women in Love*. The passage describes Ursula's confrontation of Anton under the cold moonlight of her uncle's wedding. Too long to quote in its entirety, the passage builds for several pages to the following climax:

> If he could but have her, how he would enjoy her! If he could but net her brilliant, cold salt-burning body . . . net her, capture her, hold her down, how he would enjoy her. He strove subtly, but with all his energy, to enclose her, to have her. And always she was burning and brilliant, and hard as salt, and deadly. . . . She took him in the kiss, hard her kiss seized upon him, hard and fierce and burning corrosive as the moonlight . . . cold as the moon and burning as fierce salt . . . destroying him, destroying him in the kiss. And her soul crystallized with triumph, and his soul was dissolved with agony and annihilation. So she held him there, the victim, consumed, annihilated. She had triumphed: he was not any more. (*WL,* p. 320)

Ursula's "triumph" is ironic, since it destroys the possibility for a dynamic equilibrium between them. It is this potential, even more than Anton's "core," that has been "annihilated." The struggle begins with Anton's attempt to enclose Ursula; Ursula responds by "consuming" him. Both stances imply an imbalance that would eventually lead to unmitigated linearity. An unimpeded flow of language would represent what Ursula unwittingly achieves when she destroys Anton as an independent polarity: an inherent linearity that will eventually become trapped in its own abstractions. But the strong, rhythmic pulsations of the prose help to offset this linearity, so that the language achieves what the characters cannot, an inner tension that can come to climax without being condemned to linearity as a result. If the dipolarity between the character fails, the dipolarity of the language is successful; the climax is searing in its intensity.

In this passage Lawrence manages to depict linearity without becoming condemned to it because the structure of the language preserves a sense of dipolarity that works against the triumph of linearity between the characters. Lawrence is in control of the dynamic in this passage because he allows space for both the approach to and avoidance of the "unconscious." It is when he tries to evoke one term of this ambivalence without allowing space for its contrary that he slips from paradox into contradiction or, worse, into didacticism. Given the obvious impossibility of what he hoped to accomplish, it is surprising not that Lawrence occasionally failed, but that he so often succeeded in crafting verbal illusions that are faithful to both the potential and the threat he felt in the integrating field of the "unconscious." In his most successful novel, *Women in Love,* Lawrence is able to integrate both polarities of this ambivalence into a coherent schematic that is also artistically powerful.

By the time he wrote *Women in Love,* Lawrence had so refined the dynamics of linearity and dipolarity that virtually the entire action of the novel serves to delineate their complexities. Pairs of characters come together, engage one another—on the surface through dialogue and argument, underneath the verbal surface by symbolic interplay between the unconscious of each—then move apart as they begin to experience the consequence of the dialectic they have set in motion between them. With the language insuring a continuing tension, both linearity and dipolarity can be fully explored as psychological dynamics. The to-and-fro movement has expanded to include permutations only vaguely implicit in *The Rainbow;* we are made to see more clearly that the to-and-fro can be destructive as well as synergistic. Meanwhile, the results of a relationship degenerating into linearity are also more fully represented.[15]

Within the to-and-fro motion in *Women in Love,* three possibilities emerge. The first, and happiest, possibility is that the couple will use the to-and-fro to break through to an unmediated apprehension of reality. As in *The Rainbow,* to achieve the breakthrough one must have a

[15] I will not have space here to do more than outline the general nature of the scheme, to show how it develops from *The Rainbow* and anticipates the essays. The reader who wishes more detail is referred to Howard Harper, Jr., "*Fantasia* and the Psychodynamics of *Women in Love,*" in *The Classic British Novel,* ed. Harper Edge and Charles Edge (Athens, Ga.: University of Georgia Press, 1972), pp. 203–219.

partner willing to serve as the doorway—in Lawrence's terms, a partner willing to make an irrevocable commitment to an "impersonal" union. It is thus that Ursula and Birkin serve one another. Each is reluctant at first to make the commitment, Ursula because she wants a personal love that stops with Birkin's adoration of her, Birkin because he cannot make the final break with Hermione. Each helps the other to see that the break with traditional relationships must be made. Birkin must detach himself from the "vomit" of his relationship with Hermione, and Ursula is vital in helping him finally to make that break; Ursula must not fall into the trap of merely personal love (which *The Rainbow* explores exhaustively in the relationship between Will and Anna), and Birkin is instrumental in helping her to see this. Neither Birkin nor Ursula, it seems, could make the breakthrough without the other. It is their basic complementarity and their commitment not only to each other but to the "greater reality" that allows them to engage each other in a synergistic dynamic.

There are also other, more sinister possibilities. One starts appearing after Gerald draws back from the essential commitment, first with Birkin and then with Gudrun. When Gerald and Gudrun fail to make a real commitment to each other, they become locked into a closed system, so that what fills one empties the other. Gerald comes to Gudrun in her room, pouring into her "all his pent-up darkness and corrosive death, and he was whole again" (*WL,* p. 337); but Gudrun "lay wide awake, destroyed into perfect consciousness" (*WL,* p. 338). As the conflict deepens after Birkin and Ursula leave them alone together in the Tyrol, both Gudrun and Gerald subconsciously realize that the dynamic between them condemns them to the closed economy of a system in which energy is conserved rather than generated. "Sometimes it was he who seemed strongest, whilst she was almost gone, creeping near the earth like a spent wind; sometimes it was the reverse. But always it was this eternal see-saw, one destroyed that the other might exist, one ratified because the other was annulled" (*WL,* p. 436). In the end it is Gerald who succumbs. His failure to break through to the dynamic flux of reality ultimately proves fatal, his frozen carcass symbolizing his final collapse into stasis. Though Gudrun survives, her victory is only another form of defeat. She will continue along the path of dissolution, exploring the further stages of depraved sensuality with Loerke.

The third possibility is the least firmly sketched of the three "frictional" motions; it is a to-and-fro in which the oscillations back and forth become increasingly violent, eventually leading to the permanent fragmentation of a bifurcated psyche. The emergence of this kind of motion illuminates why Lawrence should insist, in the essays, that the opposing body centers are united into a mystical whole by an interconnecting field. In retrospect the notion of a holistic field can be seen as an attempt to avoid having the to-and-fro motion degenerate into an aimless oscillation that eventually leads to dissolution of the psyche and death.

The clearest example of a to-and-fro that leads to fragmentation appears not in the published version of *Women in Love,* but in the "Prologue" that Lawrence deleted from the finished novel. In this "Prologue," Lawrence relates how Birkin shuttles between an empty sensuality and a depraved spirituality in his union with Hermione until he becomes "nothing but a series of reactions from dark to light, from light to dark, almost mechanical, without unity or meaning."[16] So damaging is this arid, meaningless to-and-fro that Birkin recognizes that he is "not very far from dissolution" ("Prologue," p. 107). The same kind of destructive oscillation appears whenever the spiritual and sensual centers are too far sundered to join even in the loose affiliation of a "dipolarity." Its principal example in the published text of *Women in Love* is the marriage of Mr. and Mrs. Crich. Mrs. Crich's mind, in reaction against her husband's Christian ideals, has become deranged so that she exists only at the animal level of the senses; Mr. Crich has kept his mind and will intact, but his body is undergoing dissolution.

To Lawrence, linearity is merely the extreme continuation of this destructive oscillation. Linearity results when one of the poles of a natural polarity is so far gone that it is altogether suppressed. Then, instead of a diverging to-and-fro that becomes an ever-widening oscillation, there is motion in one direction only. Linearity implies that the compensating, opposite movement has been altogether obliterated.

In "The Industrial Magnate" in *Women in Love,* Lawrence traces the progress of this linearity with devastating clarity. First comes the exaltation of the spiritual centers, the "ideal" in Christianity, at the expense of the dark sensual centers. Under "idealism" the natural to-and-fro mo-

[16]D. H. Lawrence, "Prologue to *Women in Love* (Unpublished)," ed. with an introduction by George H. Ford, *Texas Quarterly,* 6 (Spring 1963), 106.

tions of the body are perverted, resulting in the dominance of the conscious mind. The next stage after the Christian glorification of the ideal is the industrial worship of the machine. The two stages have in common the suppression of the sensuality that should balance conscious thought. Because the machine crushes the natural equilibrium between contraries even more brutally than did Christianity, it is the more linear, representing the next, further stage of development. Hence Mr. Crich's Christian benevolence is inevitably superseded by Gerald's efficiency. The next stage goes beyond Gerald. It is the mechanization of the body by the mind that Loerke expounds, with the pole of sensuality exploited by the contrary and now completely dominant pole of the conscious mind merely to furnish it with "sex in the head."

Lawrence's most explicit description of the process of reduction in a closed system occurs in the penultimate chapter, "Snowed Up," as Lawrence explains why Gudrun prefers Loerke to Gerald. Between any two people, Lawrence writes, "the range of pure sensational experience is limited" (*WL,* p. 443). Once these limits are reached, "there is no going on. There is only repetition possible, or the going apart of the protagonists, or the subjugating of the one will to the other, or death."

> Gerald had penetrated all the outer places of Gudrun's soul. He was to her the most crucial instance of the existing world. . . . In him she knew the world and had done with it. . . . But there were no new worlds [to conquer], there were no more *men,* there were only creatures, little, ultimate *creatures* like Loerke. The world was finished now, for her. There was only the inner, individual darkness, sensation within the ego, the obscene religious mystery of ultimate reduction, the mystic frictional activities of diabolic reducing down, disintegrating the vital organic body of life. (*WL,* p. 443)

The opposite to expansion into the infinite, then, is the reduction of the enclosure. Trapped within the finite world of sensation and conscious ideas, the "frictional activities" of engaging the other become a "diabolic reducing down," a cannibalistic feeding on the interior life of the self because one has failed to break out of the confines of self.

It will be apparent from this summary that *Women in Love* articulates the various possibilities of combining linearity and dipolarity much more precisely and fully than *The Rainbow* does. But despite this success, Lawrence is no more able than he was in *The Rainbow* to depict the ultimate goal toward which the "frictional to-and-fro" is supposedly tending—that is, the breakthrough into the creative unconscious. *Women in Love* is superior to *The Rainbow* in showing how and why the characters fail to reach this holistic reality, and in coordinating these individual failures with the larger failures of industrialized society. But it is less successful in actually rendering the experience of entering an undifferentiated reality. Ursula and Birkin's momentary breakthrough into the unconscious pales beside the fierce intensity of the relationship between Tom and Lydia in *The Rainbow*. It is dissolution and degradation that dominates *Women in Love,* not the fragile rapprochement that the married couple find.

The progression suggests that Lawrence is maturing in a very different direction than we might have predicted on the basis of *The Rainbow*. Rather than becoming more skilled at representing a holistic reality, he is becoming more adept at finding ways to represent fragmentation. Despite what Lawrence says about the "creative unconscious" being the source of true wholeness, if we judge the unconscious solely on the basis of how it appears in his work, it is an even more powerful medium of estrangement. The characteristic narrative pattern in *Women in Love* is for two characters to come together, rub each other raw so that the powerful forces of the unconscious come increasingly to dominate their actions, then break apart as the unconscious forces thus set in action begin to take their course. Very rarely—only once, in fact—does this "frictional to-and-fro" break through to the holistic reality that Lawrence celebrates in the union of Tom and Lydia Brangwen. Much more frequently the unconscious forces lead to violent antagonisms and radical bifurcation.

In the "scientific essays," we return to the inchoate ambiguities of *The Rainbow,* and with it, to a renewed, revealing tension in the relationship between child and parent. The earlier configurations persist—linearity and dipolarity, enclosure and openness—but now they are subsumed into a single term that is posited against the Freudian theory of repressed incest desire as the other polarity. In one sense Lawrence is condemned to giving scope to this hated term, lest the dialectic collapse into stasis; in another sense, Lawrence's theories take their vitality precisely from this opposition. As he struggles to articulate the crucial relationship between his theory of the unconscious and the parent-child dynamic, the underlying forces that we have been tracing in the fiction erupt into new "confusions," and the result illuminates not only Lawrence's art, but also the deeper anxieties that can accompany the encounter with undifferentiated reality.

At the heart of Lawrence's "scientific" theory is his version of infant psychology. Lawrence explains an infant's development in terms of symmetrical pairs of interacting centers. The first center to awaken in the child is the solar plexus, from which comes the infant's response toward his parents. The second center to come into play is the lumbar ganglion in the lower back, the seat of the infant's reaction away from the parents. Connected through a "polarity," plexus and ganglion interact to form a field that expresses both the need to reach out to the other and the need to experience the self in autonomy and aloneness.

According to Lawrence, the next two centers to come into play are in the upper body, through the cardiac plexus in the chest and the thoracic ganglion in the shoulders. As with the lower centers, the plexus is polarized toward the parent, whereas the ganglion is a center of volition and autonomy. But the upper centers are distinct from the lower because they operate in what Lawrence calls an "objective" mode, seeking direct knowledge of the object. The lower centers by contrast are in a subjective mode, knowing the other only through its relation to the self. With four centers in existence, there occurs the possibility of interaction along the vertical plane. Not only can the plexuses interact horizontally with the corresponding ganglia to establish a polarity, but the two ganglia can interact together, and the two plexuses. Thus incarnate in the human body are two great tensions: the polarity between the spiritual and sensual in the vertical plane; and the polarity between union and autonomy on the horizontal plane.

At puberty, Lawrence says, four more centers come into play; later in life, yet four more. The details of the scheme are less important than the overall symmetries of the resulting polarities. Two points particularly should be noticed. The first is the twofold symmetry mentioned above, the tensions between the spiritual/sensual and sympathetic/volitional. The second, entailed by the overall symmetries, is the curious fact that the genital center has its symmetrical counterpart in the upper center of the throat. The significance of this vertical symmetry will be apparent later. For the moment, I wish to consider the horizontal (sympathetic/volitional) symmetries and their relation to the "confusions" of the fiction.

In his theory of infant development, Lawrence preserves what was perhaps the central fact of his childhood: the reaction toward one parent, balanced by the reaction against the other. Significantly, Lawrence gives to the father the role of calling the volitional centers into play; the mother, he remarks, will be more likely to arouse the sympathetic centers. There is little doubt that Lawrence felt spiritually very close to his mother; but the autobiographical *Sons and Lovers* suggests that there was also considerable anxiety in that relationship, arising, the title implies, from a fear of incest. Lawrence begins *Psychoanalysis* with an attack on Freudian psychology because Freud, he believes, makes incest desire into "part of the normal sexuality of man." "Once, however, you accept the incest-craving as part of the normal sexuality of man," Lawrence writes, "you must remove all repression of incest itself. In fact, you must admit incest as you now admit sexual marriage, as a duty even" (*Fantasia,* p. 7). In Lawrence's theory, the opposing centers guarantee that an unqualified motion toward the mother is "unnatural." For every motion toward the mother, Lawrence implies, there should be an equally natural and necessary motion away. On one level, then, Lawrence's attack on Freud is a protective stategy designed explicitly to deny that unqualified attraction toward the mother is healthy or natural.

Beneath the surface, however, is a deeper protective strategy. More threatening to Lawrence than an incestuous, genital coupling with the mother is what he calls "spiritual incest," that is, a fusing of identity characteristic of the child's pre-Oedipal attachment to the mother. Freud's theory of infant psychology differs from modern developmental theories primarily in the stress that Freud puts on the Oedipal stage of development; by contrast, current work emphasizes the pre-Oedipal period. The differences in perspective are profound, because if gender identity comes not in the identification with the father but in the identification with/differentiation from the mother, identity is less a function of the Oedipal conflict than it is of the drive to differentiate oneself from the mother. When Lawrence speaks of "spiritual incest" as the malaise of our time, he anticipates the modern view, and consequently diverges significantly from Freud's theories.

Curiously, in Lawrence's essay this crucial difference is made to appear as if it were a part of, or at most an extension of, Freud's Oedipal complex. This suppression of a crucial difference is the more curious because Lawrence is very explicit, to the point of shrillness, about other differences between Freud's theories and his own ideas. Lawrence's conflation of his "spiritual incest" with Freud's Oedipal theory suggests that he is evading some recognition that distinguishing between his theory and Freud's would force him to make. By confusing the two theories, Lawrence is able to displace his anxiety about a "spiritual" union with the mother onto the less threatening (because easier to control) prospect of a genital coupling. He even names "spiritual incest" in such a way as to imply that it is merely another form of "incest desire," so that his idiosyncratic terminology serves further to conflate the two theories.

These "confusions" notwithstanding, Lawrence is not primarily concerned about "genital" incest. More fundamental for him is the child's inability to differentiate himself from the mother. The importance Lawrence places on the father as the parent who awakens the centers of resistance in the child is suggestive of the deeper strategy at work, for it is when the child engages in a relationship with the father that he moves from the earlier mother-centered stage to the phallic orientation of the Oedipal stage.

When Lawrence turns to consider "spiritual incest" explicitly, however, these strategies can no longer control the anxiety, which then erupts into new "confusions." When a woman is unfulfilled in her marriage, Lawrence writes in *Fantasia,* she turns to her son for satisfaction (just as Mrs. Morel does in *Sons and Lovers*). Concentrating all her love and sympathy on the son, the mother prematurely awakens him to adult consciousness. According to Lawrence, the child who is thus awakened will be unable to be satisfied by an appropriate mate later on, because he has become fixated on the mother. Lawrence comes close to recognizing the pre-genital nature of this attachment when he locates it not in the genital centers, but in the upper throat centers. "Spiritual incest" arises when mothers "establish a dynamic connection between the two centres, the centres of the throat, the centres of the higher dynamic sympathy and cognition. They establish that circuit. And break it if you can. Very often not even death can break it" (*Fantasia,* p. 158). This recognition is crucial, for as we shall see, it establishes a direct link between the Lawrence's "subjective science" and his art.

We have seen that Lawrence, through the "frictional to-and-fro" of his style, creates a tense and rhythmic movement of language that he likens to the contractions of labor. By imaging his speech as the act of birth, the quintessential moment when one becomes a mother, Lawrence implies that through his art he can metaphorically *become* the mother. Through this strategy, the part of one's self that seeks fusion with the mother is satisfied, for it imagines a possession that is at once more symbolic and more complete than a genital coupling could ever be, representing not merely possession of the body but appropriation of the essence.

We can now also understand why the genital and throat centers are connected in Lawrence's theory. The displacement of genital bonding by bonding through the throat is appropriate (and even necessary) as a reinforcement to the artist-becoming-the-mother, for the throat is, of course, the place from which speech issues. Readers of *Sons and Lovers* have long recognized that the tie with the mother is one of the deep springs of Lawrence's art. The "scientific" essays confirm and extend this insight by showing how, for Lawrence, the yearning to possess the mother is symbolically transformed, in a rich alchemy that produced the early novels, into the need for artistic speech.

How then are these strategies related to the doorways and enclosures that are central metaphors in Lawrence's fiction? When Lawrence re-creates the quintessential moment of becoming the mother in his art-speech, he is simultaneously appropriating for himself the mother's power of creation and freeing himself from her dominance over him; his art defines him as an artist with the power of creation, rather than as a son who is the creation of his mother. But since he has also in some sense become the mother, liberation and dependence are deeply entwined. Hence the characteristic conflation of enclosures and doorways, and their archetypal expression as wombs and vaginas. When the power struggles between male and female characters in Lawrence's early fiction are most intense, they invariably erupt into these images. Recall that Anton, in the long passage quoted earlier from *The Rainbow,* is trying to "net" Ursula, to "enclose" and "capture" her. Ursula's corrosive energy, too strong for Anton, breaks through these bonds and instead captures his "core." Thereafter he is her creature, something she has "created." What we are witnessing in this scene is the reverse of a birth, the regression of Anton from the independence of the adult to the complete dependence of a child in the womb. This is the ultimate horror that Lawrence through his art-speech can re-create and, through the act of re-creation, also escape, although he never entirely loses the sense that the freedom and enclosure, escape and capture, are two alter faces of the same spectre.

Lawrence's ingenuity in transforming "spiritual incest" from a confining enclosure to the openness of the creative act is matched by his honesty in admitting that the transmutation is only partially effective. When he wrestles explicitly with the "ghoul" of incest desire, the strategies of transformation are stressed to the breaking point. One such point of stress is the passage in *Fantasia* on the interpretation of incest dreams. "It is *always* wrong," Lawrence writes, "to accept a dream-meaning at its face value."

> Sleep is the time when we are given over to the automatic processes of the inanimate universe. Let us not forget this. . . . In the case of the boy who dreams of his mother, we have the aroused but unattached sex plunging in sleep. . . . We have the image of the mother, the dynamic emotional image. And the automatism of the dream-process immediately unites the sex-sensation to the great stock image, and produces an incest dream. But does this prove a repressed incest desire? On the contrary. (*Fantasia*, pp. 196–197)

In this argument, Lawrence sets up terms which, if extrapolated to their obvious end point, lead to the reasonable conclusion that incest dreams are an indication of incest desire. But then he reverses the line of argument to say that the opposite conclusion is true. As Lawrence struggles to keep his defenses intact, the language becomes highly stressed. "The truth is," Lawrence writes, "every man has, the moment he awakes, a hatred of his [incest] dream, and a great desire to be free of the dream, free of the persistent mother-image or sister-image of the dream. It is a ghoul, it haunts his dreams, this image, with its hateful conclusion" (*Fantasia*, p. 197). Even as he writes this painful truth, Lawrence is trying to escape the "hateful conclusion" by another abrupt shift in direction. The actual cause of the incest dream, Lawrence says, is not the mother but the wife. Then, again, the painful return, the implicit recognition: "But even though the actual subject of the dream is the wife, still, over and over again, for years, the dream-process will persist in substituting the mother image. It haunts and terrifies a man" (*Fantasia*, p. 197).

In these and surrounding passages, linearity and incest come together with the image of the machine. The dream image of the mother, Lawrence says, "refers only to the upper plane." When the "automatic logic" of the dream unites this upper image with lower genital desire, it is not an authentic connection but a "piece of sheer automatic logic." To proceed in a straight line is to act like a machine, not a living being. Because life is not linear, the linear logic of the incest-dream that connects the upper spiritual desire with genital lust only proves that a man's living soul could not be implicated in this mechanical conclusion. Lawrence thus derives the contradictory result that an incest dream proves not the presence of incest desire but the "living fear of the automatic conclusion." The mother image, Lawrence writes,

> was the first great emotional image to be introduced into the psyche. The dream-process mechanically reproduces its stock image the moment the intense sympathy-emotion is aroused . . . the mother-image refers only to the upper plane. But the dream-process is mechanical in its logic. Because the mother-image refers to the great dynamic stress in the upper plane, therefore it refers to the great dynamic stress in the lower. This is a piece of sheer automatic logic. The living soul is *not* automatic, and automatic logic does not apply to it . . . the living soul *fears* the automatically logical conclusion of incest. (*Fantasia*, p. 198)

The reasoning in these passages is so tense, so nonlinear, in a sense, that even Lawrence admits his argument "may sound like casuistry." If it is casuistry, however, it is no less significant because of that. To proceed linearly means that one remains trapped within a lifelong desire for the mother; it also means that one acts not like a person but like a machine. One can avoid this linearity by speaking, by using language in a to-and-fro dialectic that has the power to break free of enclosures. The nexus between linearity, the machine, enclosure, and incest thus emerges in a way that allows us to see how it could be connected with Lawrence's need to speak, as well as with the highly stressed metaphoric patterns in the art speech.

Though Lawrence's art remains most powerful in its rendering of an ambivalent approach to/avoidance of the undifferentiated field of the unconscious rather than an entrance into it, at its best it achieves a remarkable transformation of private concerns into patterns of universal significance. The universality of this dynamic can be appreciated when we see it in a discipline as far afield from Lawrence as quantum mechanics. Lawrence shares with quantum physicists an early attachment to the mother as primary caretaker, and a process of gender differentiation in which the mother is posited as quintessentially different from the self. Thus for both artist and scientist a subconscious equation is made between the autonomy of the self and the objectification of the other. For the scientist, this early objectification of the mother is transmuted into a later objectification of Mother Nature, whereas for Lawrence it is changed into the complex tensions of his fiction and essays. So when Lawrence imagines in his art an integrated unconscious "field" in which the boundaries between self and other are blurred, he is exploring the same kind of anxiety that a quantum physicist feels in the presence of a scientific model that similarly blurs the boundaries between subject and object. For both, the anxiety is linked with the deepest levels of identity and is bound up with the drive to achieve an autonomous identity independent of the mother.

Lawrence's art, then, though it is based in part in his own personal relationships with women, transcends the merely personal. At its best, it is a complex and sometimes tortured exploration of the anxieties and ambivalences that can occur whenever one encounters a field model of reality, whether in fiction or in quantum mechanics. The irony is that Lawrence himself was almost completely unaware of these parallels. He saw in modern science merely the tendency to objectify reality, without realizing that it too was undergoing a radical transformation in the face of modern field theories. The larger terms of Lawrence's dissolving dialectic are thus Lawrence and the science he did not understand. Believing that the most fruitful interaction comes not from consensus but from passionate struggle between contraries, Lawrence made science into the essential other term necessary to begin the dialectic, thereby transforming his sense of alienation from it into an asset. If Lawrence's theories are finally mystical rather than scientific, intuitive rather than logical, they assign to science the essential role of the "other." It is fitting that this binary opposition itself blends into a complicated field of the kind that Lawrence reached out toward, but could never entirely grasp.

CHAPTER 4

AMBIVALENCE

Symmetry, Asymmetry, and the Physics of Time Reversal in Nabokov's *Ada*

As far as the laws of mathematics refer to reality, they are not certain; and as far as they are certain, they do not refer to reality.

Albert Einstein, *Ideas and Opinions*

IN *Ada* Nabokov interprets the field concept with considerable license, using the scientific models more as catalysts for his own ideas than as well-defined paradigms he follows. In this he is like Lawrence; he is like him too in investing his literary strategies with intense ambivalence. But the roots of the ambivalence are essentially different in the two writers. As we have seen, Lawrence was strongly attracted to a field concept because of the union between subject and object that it seemed to promise; he was, however, also wary of the loss of individuation that this fusion could entail. Nabokov, by contrast, is relatively uninterested in the mystical union of self with other that so fascinated Lawrence. A more cerebral writer than Lawrence, Nabokov is also more adamant about preserving the boundaries of the ego intact. For him the attraction of quantum field theory lies in the possibility implied by its broken symmetries (for reasons we will discuss shortly) that time is reversible. Nabokov can accept this possibility only at the price, however, of subjugating the freedom of his artistic creation to the limitations that he associates with scientific observation. The central issue in *Ada* is therefore not autonomy but control, and it is rendered not through the polarities that Lawrence adapted from Maxwellian electrodynamics, but through the mirror symmetries that dominate both Nabokov's fictions and the unified field theories of modern physics.

Symmetry is fundamental to field theory because it is through its symmetrical properties that the field is described. When the world is conceived atomistically, as a collection of material points arranged in space, the symmetry of any given arrangement is an accidental property of the system. But with the shift to the field as "the only reality," the symmetries of the underlying field become the chief means by which particle interactions are understood and predicted. The shift of emphasis from an atomistic to a field concept thus transforms symmetry from an accidental to an intrinsic property, and consequently places symmetry considerations at the center of modern physics. Werner Heisenberg describes in his memoirs how he and his colleague Wolfgang Pauli came to see symmetry as the key to a unified field theory. "'In the beginning was the symmetry' is certainly a better expression than Democritus' 'In the beginning was the particle,'" Heisenberg asserts.[1] The importance of symmetry to field theory has been underscored recently because of a series of experiments indicating that the symmetries of the underlying field are not universally upheld. These violated or "broken" symmetries have led to renewed speculation about the role of symmetry in field theory. To understand them, we shall need briefly to review what symmetry operations are, and how they enter into particle physics.

Perhaps the most familiar kind of symmetry operation is the reflection of an object in a mirror. If the reflected image can in theory be taken out of the mirror and superimposed on the object, then the object is said to possess mirror symmetry. There are also other kinds of symmetry operations; for example, if an object, after being rotated n degrees still looks the same, it possesses n-rotational symmetry. The three symmetries of most interest to particle physicists are charge symmetry, in which positive and negative charges can be interchanged; parity, the equivalence of left- and right-hand mirror images; and time symmetry, in which $-t$ can be substituted for t in the field equations without violating any known laws. Until the mid-fifties, all three of these symmetries were thought to obtain throughout nature. In 1956, however, C. S. Wu and her associates, in a historic experiment, proved that

[1]Werner Heisenberg, *Physics and Beyond: Encounters and Conversations* (New York: Harper & Row, 1971), p. 240.

electrons were emitted preferentially upward (that is, in the direction of the magnetic field) in the decay of radioactive cobalt.[2] This result proved that parity was not upheld, for had parity obtained, the electrons should have been emitted equally up and down ("up" and "down" in this context can be thought of as equivalent to "right" and "left"). As Martin Gardner explains in *Scientific American,* this meant that "there are events on the particle level . . . that cannot occur in mirror-reflected form."[3]

Physicists sought to extricate themselves from this uncomfortable situation by postulating that asymmetries in charge could cancel out the asymmetries in parity. Thus, if charge and parity were considered together as a single CP symmetry, the CP symmetry could hold even though the symmetry of neither P nor C by itself could. In terms of Wu's experiment, CP symmetry would be conserved if there existed a parity- and charge-reversed theoretical counterpart to cobalt, cobalt made of antimatter or "anti-cobalt," which emitted electrons preferentially downward. Then nature could still be said not to have a preference for one "hand" over the other, because the two cobalt emissions, one up and one down, would in theory cancel each other out.

But the preservation of CP symmetry was in turn thrown into doubt by a 1964 experiment on the decay of K mesons which implied that CP symmetry was violated. Now the only way for physicists to salvage the overall symmetry was to assume that asymmetries in time could cancel out the asymmetries in charge and parity, so that even if CP symmetry did not hold, CPT symmetry would. But this in turn implied that time-reversal symmetry did not hold by itself, so that the last of the three single symmetries was also assumed to be violated.

Despite the steady encroachment of asymmetries into the field model, physicists regard it as extremely unlikely that the overall CPT symmetry will fall. Eugene Wigner, a seminal researcher in this area, explains why: not because physicists love symmetry (though some do), but because of the "stubborn fact that we cannot formulate equations of motion in quantum field theory that lack this symmetry and still

[2]A full account of these developments is given by T. D. Lee, "Space Inversion, Time Reversal and Particle-Antiparticle Conjugation," *Physics Today,* 19 (March, 1966), 23–31.
[3]Martin Gardner, "Can Time Go Backward?" *Scientific American,* 216 (January 1967), 200.

satisfy the postulates of Einstein's special theory of relativity."[4] Hence, if CPT symmetry should not hold, a revision of the very foundation of modern field theory would be necessary.

Even if overall CPT symmetry is conserved, however, the collapse of the individual symmetries has implications that Wigner finds disturbing. Because there is at present no theoretical explanation for why nature should prefer one "hand" over the other, physicists are forced to conclude, Wigner points out, that although "two absolutely equally simple laws of nature are conceivable, nature has chosen, in its grand arbitrariness, only one."[5] Thus, though a universe in which the CP symmetry tilts one way is as conceivable as one in which it tilts the other, one is consistent with the laws of nature, the other not.

As we shall see, there is an extraordinary congruence between these scientific developments and Nabokov's conception for *Ada*. The congruence can be traced certainly to one source, Martin Gardner's *The Ambidextrous Universe* (first published in 1964), and less certainly to Gardner's 1966 *Scientific American* article on time reversal ("Can Time Go Backward?"). Published in 1969, *Ada* as a novel apparently grew out of the shorter philosophical work, *The Texture of Time,* that Nabokov identifies in a 1966 interview with Alfred Appel as its "central rose-web."[6] Nabokov would claim in 1970 that, "whatever I may have said in an old interview," the speculations on time apply only to Part Four of *Ada,* not to the entire novel; nevertheless, it seems clear from the "time-wrenched" cosmology of *Ada* that Van's physical and metaphysical inquiries into time are central to its conception.[7] It is also clear from internal evidence that Nabokov was familiar with Gardner's book. In the first edition of *The Ambidextrous Universe,* Gardner had quoted lines from the poem "Pale Fire," puckishly attributing them to the "poet John Shade" without mentioning Nabokov. In *Ada* Nabokov quotes the same two lines, attributing the quotation to an "invented philoso-

[4]Eugene P. Wigner, "Violations of Symmetry in Physics," *Scientific American,* 213 (December 1965), 34.

[5]Ibid., p. 36.

[6]Vladimir Nabokov in "An Interview with Vladimir Nabokov," conducted by Alfred Appel, Jr., in *Nabokov: The Man and His Work,* ed. L. S. Dembo (Madison: University of Wisconsin Press, 1967), p. 43.

[7]Vladimir Nabokov, "Anniversary Notes," Supplement to *TriQuarterly,* 17 (Winter 1970), p. 5.

pher," "Martin Gardiner."[8] The correspondence between the two works is, however, far more extensive than this polite exchange of pleasantries.

Gardner's book examines the extensive role that mirror symmetry plays in life on earth, from man's bilateral symmetry to the double helix of DNA.[9] The final chapters deal with the "Ozma Problem," the question of whether there is any way to describe the difference between right and left in absolute terms, without referring to other conventions. The problem is usually posed as how to communicate what right and left mean to Planet X, assuming that only words, not pictures or common reference points, may be transmitted between the two planets. The question the Ozma Problem asks is this: are the two halves of left-right symmetry as they are found in fundamental structures exactly equal? Or is there some asymmetry that allows us to distinguish between them? In short, is the universe ambidextrous? Gardner explains that the Ozma Problem was answered in the discovery that parity is not conserved. The fall of parity has implied, of course, that a slight skew exists in nature, a slight preference for one "hand" over the other.

Gardner goes on to discuss what implications the fall of parity has for the other major symmetries of charge and time. The most important is that antimatter, or more precisely antiparticles, exist. Although the first edition went to press before it was discovered that CP symmetry also may not hold, Gardner nevertheless links the fall of parity with time reversal by suggesting that an antiparticle could be an ordinary particle that has been rotated through a higher dimension, for example through the "fourth dimension" of time. Through this reasoning Gardner anticipates his later *Scientific American* article explaining why the symmetry violations of particle physics imply that time is reversible.

In his *Scientific American* article, Gardner deals explicitly with whether time can be reversed and, if so, what it means to say a world is moving backward in time. He points out that it makes sense to say time goes "backward" only if we ourselves are moving in the opposite direction; otherwise, all we would be able to know is that time moves.

[8]Vladimir Nabokov, *Ada or Arbor: A Family Chronicle* (New York: McGraw-Hill, 1969), p. 577.

[9]Martin Gardner, *The Ambidextrous Universe* (New York: Charles Scribner's Sons, 1964).

Gardner thus asserts that "it is only when *part* of the cosmos is time-reversed in relation to another part that such a reversal acquires meaning." He then discusses some of the reasons why we would not be able to communicate with a time-reversed world. "If you somehow succeeded in communicating something to someone in a time-reversed world," Gardner explains, "he would promptly forget it because the event would instantly become part of his future rather than his past."[10] The occasion for the article was, of course, the recently discovered symmetry violations in high-energy physics, so that time-reversal was placed in a context that linked it with a slight asymmetry in nature's generally symmetrical design.

Even without the more explicit arguments about time reversal in Gardner's *Scientific American* article, enough is said about it in the first edition of *The Ambidextrous Universe* to serve as a powerful stimulus to Nabokov's imagination. Gardner himself, in the revised second edition of *The Ambidextrous Universe* (published in 1979), remarks upon the parallels between his book and Nabokov's 1974 novel *Look at the Harlequins!* Gardner modestly remarks that "questions about the symmetries of space and time are so essential to the plot that I like to think that the book was influenced by Nabokov's reading of the first edition of this book."[11] A reader less tied to the demands of modesty would see in Gardner's book not merely an influence, but a seminal conception that Nabokov appropriated for his own purposes.[12]

Ada is the pivotal text for Nabokov's new conception of symmetry. Gardner had suggested there could be no communication between us and a world moving backward in time; in *Ada* Nabokov imagines that there can be no direct communication between his two "time-wrenched" planets, Terra and Antiterra. Gardner recounts how physicists resisted the intrusion of asymmetry into their field theories; Nabokov creates a protagonist who desperately searches for symmetry but discovers instead the slight asymmetries that defeat his expectations. Gardner writes about

[10]Gardner, "Can Time Go Backward?" p. 102.

[11]Martin Gardner, *The Ambidextrous Universe: Mirror Asymmetry and Time-Reversed Worlds,* 2d ed. (New York: Charles Scribner's Sons, 1979), p. 271.

[12]The argument for the link between Gardner's *Ambidextrous Universe* and *Look at the Harlequins!* has been made by D. Barton Johnson in "The Ambidextrous Universe of Nabokov's *Look at the Harlequins!*" in *Critical Essays on Vladimir Nabokov,* ed. Phyllis Roth (Boston: G. K. Hall, 1984). I am grateful to Professor Johnson for sharing his work with me prior to its publication, and for his suggestions for this chapter.

scientific experiments that imply time can run backward; Nabokov makes Van a scientist, and has him devote his later life to a treatise suggesting that time can repeat itself. And in *Ada* as in Gardner, symmetry and asymmetry are deeply bound up with the question of whether time can be reversed. The connections are implicit in the way Van structures his narrative.

In arranging his material, Van chooses to emphasize the repetitions of patterns he first encountered in the summer of 1884 when he fell in love with Ada. The repetitions suggest that time can be made to repeat itself, for no matter what chronological time has passed, in Van's "Real Time" the same events keep repeating themselves in varying configurations. The constellation that began the repetitions (the coming together of Van and Ada) was itself composed of the joining of two reflective images. Ada's "right instep and the back of her left hand" bear the same "indelible and sacred birthmark" that marks Van's right hand and left foot (p. 230), Ada's "plain Irish nose [is] Van's in miniature" (p. 64), and her hands are "Van's in a reducing mirror' (p. 403). Because Van, in possessing Ada, is co-joining mirror images out of which the later reflections evolve, he is always anxious whenever she changes; change threatens the mutuality of reflection. But the mirror correspondence between them miraculously continues as they grow older. In middle age Van sees that they have had the same molar, though on opposite sides of their mouths, drilled and filled with gold.

However, Van is not always so fortunate. Though he aims for exact mirror correspondence, frequently all he can achieve is a reflection that has been slightly wrenched from the original. When these asymmetries intrude upon him, Van is forced to recognize that exact repetition of earlier events is not possible, and hence he is confronted with the truth that all things change in time. The displacement from exact reflection is therefore nearly always invested with connotations of failure, because it implies time does pass; conversely, the successful replication of image brings intense satisfaction, because it implies that the past can be recaptured. The contrast is apparent in Van's unqualified ardor for Ada, in which symmetry is confirmed, and his ambiguous relationship with Lucette, into which some slight asymmetry inevitably intrudes. Lucette is a wrenched image of Ada, and broken symmetry between the two ultimately proves tragic.

Ada's colors are black and white, often modulating into black and yellow. Ada has black hair and white skin; the divan on which she and Van first make love is black with yellow cushions; she wears yellow slacks and a black bolero on the day Van learns of her infidelities to him. Lucette's color, on the other hand, is red. She has russet-colored hair, and though strongly resembling Ada, repeats the image in a different tone. At one point Lucette propositions Van by mentioning that she too has a black divan and yellow cushions. But Lucette fails because she can never duplicate Ada exactly. Her crime, Van thinks, "was to be suffused with the phantasmata of the other's [Ada's] innumerable lips" (p. 400) while never *being* Ada. Lucette finally commits suicide wearing, in an inversion of Ada's colors, black slacks and a lemon blouse. Van, riding back from the family picnic in 1888 with Lucette on his lap, remembers the occasion four years earlier when Ada rode there: "Family smell; yes, coincidence; a set of coincidences slightly displaced; the artistry of asymmetry . . . but it was that other picnic which he now relived and it was Ada's soft haunches which he now held as if she were present in duplicate, in two different color prints" (p. 296).

As Van tells his story, then, two conflicting principles compete in the organization of the narrative: the desire to create exact reflectsion, and the frequent wrenching of these into slightly displaced variations. One of the mirror-reflections Van creates is Mascodagama, the version of himself when he dances on his hands. Van likens the pleasure he takes in his Mascodagama act to the later convolutions of his writing: "It was the standing of a metaphor on its head not for the sake of the trick's difficulty, but in order to perceive an ascending waterfall or a sunrise in reverse: a triumph, in a sense, over the ardis of time" (p. 197). Lucette's governess explains to her that in Greek, "ardis" means "the point of the arrow," and it is the linearity of time's arrow, the uniform direction of its flight, that Van sees defeated in the circularity of mirror reflections. Symmetry represents Van's triumph over time; asymmetry, the inevitable admission that time passes, people change, people die. It is when he is forced to accept the asymmetries that Van feels most desperate about maintaining his control over time. The issue of control is central, then, both to Van's attitude toward time and to the tension between symmetry and asymmetry that runs through the text.

To understand how the issue of control relates not only to Van's attitude toward time, but also to his artistic arrangement of his material, we may consider how ordinary time is measured. Take for example a ticking metronome. The metronome works by holding all variables but one constant: the sound of the tick, the intervals between ticks, the length of the tick and so on, are always the same, infinitely replicable. Despite the overwhelming sameness, we are aware that the second tick is an event distinct from the first. When we try to define the difference that makes the events discrete, we are left with time. The events, identical in every other respect, are different because they take place at different moments. Unvarying similitude allows us to concentrate on the defining difference, and hence to approach pure time—that is, pure time in the ordinary sense.

In *The Texture of Time,* Van tries to free time from this arbitrary sequence and make it a function of human perception by asserting that "real Time" (denoted by a capital "T") is *not* a series of identical moments, but a sequence of events that can be either extended or compressed, depending on whether the alert, "tense-willed" mind attends to them or not. Through this argument, Van hopes to establish the primacy of human imagination over time. But because he sacrifices, in the process, the similitude inherent in clock time, he therefore risks making it impossible to measure Time at all, since measurement of time depends upon a uniform series of events against which it may be discerned. So he attempts to replenish the similitude from his own craft, creating uniformity among events through extensive repetitions and reflections. Since the similitudes come from the narrator's craft in organizing his material, they express his will (express, in fact, his obsession with Ada), in contrast to the similitude inherent in ordinary time which derives from a linearity indifferent to human desire. The repetitions in *Ada* can therefore be seen as a strategy to defeat the linearity of time, and in a sense to humanize it.

Van's ultimate nightmare is to be caught in a world that refuses to reify the subjective patterns of his thought. After Demon discovers the affair between Van and Ada and forces them to break it off, Van experiences the horror of linearity: "Numbers and rows and series—the nightmare and malediction harrowing pure thought and pure time—seemed bent on mechanizing his mind" (p. 478). The intricate patterns of repetition in *Ada* are an attempt to break that linearity, to force it to conform to the subjective patterns of time arising from the narrator's own preoccupations.

Moreover, if we are to understand Van's narrative, we are obliged to enter into those rhythms, and hence to share his preoccupations. Consider how a text such as *Ada* must be read if it is to be understood. Because so many passages reflect or vary details from previous passages, the reader who fails to remember these superabundant details will find subsequent passages increasingly unintelligible. We realize that we must not only closely attend to the present details, but retain those details as they move into the past in order to understand the future details moving into our present. Understanding the text, even on a literal level, thus requires that we duplicate Van's "tense-willed" mind and his dedication to accurate recall.

Furthermore, by entering into this dynamic, we are also valorizing Van's philosophical speculations on the nature of time. As the memory-patterns slowly accumulate in our minds through interlocking details, both present and future are put in the service of the past—precisely what Van attempts to do philosophically in his *Texture of Time*. For as we continue to read, we recognize the present as a repetition of or variation on the past, and therefore organize it *in terms of* the past. The accumulation of past detail in the reader's mind is what makes the present detail memorable. As the patterns accumulate in memory, more and more of the present is organized in those terms. Van writes, "What we do at best (at worst we perform trivial tricks) when postulating the future, is to expand enormously the specious present causing it to permeate any amount of time with all manner of information, anticipation and precognition" (pp. 596–597). The repeated reflections and their associated details have just the effect that Van describes, causing the "specious present" to expand until we share Van's perception that he lives his life out of the single summer of 1884 when he first fell in love with Ada. We, no less than he, are forced into a nostalgic stance if the present or future is to have any meaning. As we read through *Ada* the past is constantly expanding as an organizing principle, while the unknown, unforeseeable potential that we call the future is contracting.

Recall that the measurement of time depends upon both similitude and difference. We have seen how Van organizes his narrative to emphasize the repetitions, thus supplying similitude between events and linking them together in time. But unvarying reflection, being circular, would obliterate time. Making us aware that time has passed are the displacements from exact reflection, which Van always to some extent resists. The symmetries serve the purpose of defeating the ordinary linearity of time; the asymmetries allow us to distinguish between similitudes and hence to define them as discrete events separated in time.

Van recognizes, at least intellectually, the danger implicit in the nostalgic repetitions that threaten to proliferate endlessly into the future. In his treatment of Time, Van admits that a completely "determinate scheme" would create a predictable unfolding as pernicious as the clock-time that stretches toward an exactly divided, and hence infinitely predictable, future. "The determinate scheme," Van writes, "would abolish the very notion of time. . . . The determinate scheme by stripping the sunrise of its surprise would erase all sunrays" (p. 597). What Van does not seem fully to realize, however, is how closely related the perils of a "determinate scheme" are to the dominance of the past that he so much desires.

To escape the dangers of an overdetermining past, Van in his theory of Time turns to the future. He writes that "the future remains aloof from our fancies and feelings. At every moment it is an infinity of branching possibilities" (p. 597). However true this is for Van as he lived his life (and one wonders how true it can be, since Van perceives his life as being lived out of moments in the past), it is less true when we consider how Van treats time in his narrative. Van, always teetering on the edge of a "determinate scheme" dictated by the past, attempts to break out of it into the less constrictive time of his present by injecting that present into the text, for example through the notes that the present Ada and Van exchange throughout their reminiscences of the past. He thus creates an artificial future, as it were, for the reader in the text.

But this does not alleviate the problem, since that "future" (that is, Van's present time as writer of the memoir) is seen by Van as the culmination of the patterns originating from the past. For example, Van begins the chapter in which the incestuous affair between Van and Ada is discovered by announcing that the many precautions they took were "all absolutely useless, for nothing can change the end (written and filed away) of the present chapter" (p. 458). By injecting his present into the past, Van has not so much redeemed his narrative from overdetermination as he has extended the dominance of the past even further, into a future which is becoming for the reader as determined as the past.

So Van tries other ploys to relieve the sterility of a "determinate scheme." He denies, for example, that there is *the* future, only *a* future, by introducing time forks into the narrative, problematic futures that exist but that are not taken in *this* unfolding of the narrative. But here we arrive at a moot point: is *Van* introducing these time forks? Or is he merely describing a feature of the fictive world that Nabokov has created? With the time forks, the answer is unclear; they could be either one, a trick of the narrator or a real feature of Van's world. But there are other points at which the events deviate from Van's predicted pattern and thereby fail to conform to his subjective desires—points, that is, where he is forced to recognize the asymmetries. Whatever may be Van's appreciation of the dangers of a "determinate scheme," the countermeasures he takes as an artist are ineffective in escaping them. The real deviations from exact reflection are beyond Van's control, often bitterly resisted by him and proof of the partial failure of his attempt to establish the dominance of patterns emerging from the past. They arise not from his will as narrator, but from the will of *his* creator, Nabokov.

Thus, it is Nabokov rather than Van who takes seriously the dangers in a "determinate scheme." Because Van gives his allegiance to science as well as art, his control over his material is only partial; he can arrange, he can interpret, but he is not free altogether to invent the events he narrates. We as readers can perceive how, despite Van's frantic efforts, these events fail to conform exactly to the symmetries he wishes them to embody. Ada's infidelities, the repeated and painful separations from her, the desperate search for her reflection in the pitiful girl-prostitutes, all testify to Van's inability to bring exact reflections into being through the exercise of his will.

Yet at the same time, Nabokov permits Van's endeavor to be mostly successful. There is, after all, an astonishing abundance of mirror correspondences in Van's world, from the physical similarities between Van and Ada to the cosmological link between Terra and Antiterra. Nabokov and Van collaborate in establishing the dominance of the past as an organizing principle. They differ in their desire to create unvarying similitudes. Van would have complete replication, but he is denied this by his Fate—that is to say, by Nabokov. Nabokov is engaged, then, in a delicate balancing act. On the one hand he creates extensive patterns through reflections, doubling, and repetitions, thus putting the present and future in the service of the past, from whence the patterns originate and accumulate. On the other hand, he strives to escape the determinacy of over-patterning through displacements from exact reflections. Not surprisingly, these displacements are heavily invested with ambivalence, since they are at once a threat and promise. Because they vary the pattern, they contain an implicit threat to it as an organizing principle (and hence to the dominance of the past which is at the very center of *Ada*). At the same time, however, they also promise to liberate the future from the tyranny of an over-determining past.

We are now in a position to understand better the role that Terra and Antiterra play in the novel's larger patterns. The interplay between the two worlds is an expression, on a cosmic scale, of the tensions in Nabokov's artistry. The two worlds mostly mirror one another; but there are significant departures from exact reflection. There is, for example, a time gap of anywhere from fifty to a hundred years between the two worlds, so that they are separated in time as well as in space. Moreover, even given the time difference, the reflections between the two worlds are not exact, with "not *all* the no-longers of one world corresponding to the not-yets of the other" (p. 20). Terra is therefore not an exact mirror of Antiterra, but a "distortive glass of our distortive glebe," as one Antiterran scholar put it (p. 20).

Because wrenched reflections are always regarded ambivalently in this text, the moral values attached to these two slightly asymmetric worlds are ambiguous. After Lucette's suicide, Van imagines how much better her life might have been had she been on Terra rather than on that "pellet of muck," Antiterra. Judging by Antiterra's proper name, Demonia, we might suppose that this judgment is correct—until we remember that on that mirror planet demons are "noble iridescent creatures with translucent talons and mightily beating wings" (p. 23). Van is disenchanted with Antiterra in times of crises, seeing it as a "multicolored and evil world" (p. 319); but in cooler moments he writes *Letters from Terra,* proposing that the "strain of sweet happiness" that Demonians suppose they hear from their sister-planet Terra is a fraud. The "purpose of the novel was to suggest that Terra cheated, that all was not paradise there, that perhaps in some ways human minds and human flesh underwent on that sibling planet worse torments than on our own much maligned Demonia" (p. 363). Whether Terra is Heaven or Hell, and consequently whether the fictional Antiterra is a twisted parody of its heavenly Terran counterpart or a paradisical twin to our own "pellet of muck," is left ambiguous. All that we can say with certainty is that neither position is left unqualified. Given the terms in which Nabokov creates symmetries and asymmetries in *Ada*, we might almost say that ambivalences arise whenever mirror images are posited; they are a necessary consequence of cosmic doubleness.[13]

The fascination with broken symmetries that we see in *Ada* has deep roots in Nabokov's thought. Mirror symmetry, the circularity of mutual reflection, is generally expressed by Nabokov as a conviction that all true things are round. In the Appel interview, Nabokov insists at one point that "a real good head . . . is round."[14] Carol Williams, in her article "Nabokov's Dialectical Structure," uses as her epigraph these lines from Nabokov's poem "An Evening of Russian Poetry":

> Not only rainbows—every line is bent, and skulls
> and seeds and all good worlds are round.

Williams believes these lines "contain the essence of Vladimir Nabokov's metaphysical division. The human eye, he implies, can see only half of the circle (the rainbow's arc); the other half must be taken on faith."[15] But the roundness, Williams quickly points out, is not exactly the roundness of a circle—more precisely, it is the circle "set free" in a spiral. "I thought this up when I was a schoolboy," Nabokov writes in *Speak, Memory,*

> and I also discovered that Hegel's triadic series . . . expressed merely the essential spirality of all things in their relation to time. . . . If we consider the simplest spiral, three stages may be distinguished in it, corresponding to those of the triad: We can call "thetic" the small curve or arc that initiates the convolution centrally; "antithetic" the larger arc that faces the first in the process of continuing it; and "synthetic" the still ampler arc that continues the second while following the first along the outer side.[16]

[13]The ambivalence may surface in another way in the incest motif that increasingly seems to occupy Nabokov in his late English fiction, especially in *Ada* and *Look at the Harlequins!* An incestuous coupling simultaneously violates and achieves difference; on the one hand it violates socially decreed kinship differences, but at the same time it catapults the offenders outside the social norm, thereby insuring their difference within their society. These ideas were suggested to me by my reading of D. Barton Johnson's "The Labyrinth of Incest in Nabokov's *Ada*" (forthcoming in *Contemporary Literature*), which makes clear the extent of the incest theme in *Ada*, not only in the treatment of Van and Ada but also in that of their ancestors in the Zenski-Veen family line.

[14]"An Interview with Vladimir Nabokov," p. 33.

[15]Carol T. Williams, "Nabokov's Dialectical Structure," in Dembo, p. 165.

[16]Vladimir Nabokov, *Speak, Memory: A Memoir* (New York: G. P. Putnam's Sons, 1966), quoted in Williams, p. 165.

The spiral, then, accommodates the demands of symmetry, yet at the same time allows for a slight asymmetry that, in twisting the spiral upward, liberates the return from the prison of a closed circle.

In Nabokov's work as a whole, asymmetry is linked, L. S. Dembo suggests, with "the artistic need for escape from necessity."[17] It is what keeps the return from vicious circularity. Dembo quotes from Cherdyntsev, the "pure" artist of *The Gift:*

> The theory I find most tempting [is] that there is no time, that everything is the present situated like a radiance outside our blindness. . . . And if one adds to this that nature was seeing double when she created us . . . that symmetry in the structure of live bodies is a consequence of the rotation of worlds . . . in our strain toward asymmetry, toward inequality, I can detect a howl for genuine freedom, an urge to break out of the circle.[18]

Here we have an interpretation of asymmetry that is opposite to the value Van gives it in *Ada.* To Cherdyntsev it implies freedom, whereas Van laments it as a regrettable accident that interferes with the desired circularity of time. "The irreversibility of Time," Van writes, "is a very parochial affair: had our organs and orgitrons not been asymmetrical, our view of Time might have been amphitheatric and grand" (p. 573).

But what the narrator construes as his defeats—the asymmetries that impose themselves upon him—are also the very qualities that rescue his world from unreality. From this point of view, Van Veen is perhaps the most favored of Nabokov's narrators. The asymmetries that Cherdyntsev as artist must *create,* Nabokov weaves into the fabric of the universe in *Ada.* All that Van has to do is recognize them. Hence he is a scientist as well as an artist: his dedication to accurate observation holds in check his tendency to create the world, balancing the artificial world of artistic creation against the recognition of a natural world in which partial failure is inevitable. Significantly, Van is the one Nabokovian hero who finally achieves his desired end in something like the fullness and serenity he imagined. The limitations that Nabokov imposes upon his protagonist in *Ada* do not puncture the world of artistic creation, but collaborate with it to redeem it from its own excesses.

[17]L. S. Dembo, p. 15 ("Vladimir Nobokov: An Introduction").

[18]Vladimir Nabokov, *The Gift* (New York: Popular Library, 1963), pp. 384–385, quoted in Dembo, p. 15.

In this sense *Ada* is the inverse of *Pale Fire,* the novel that immediately preceded it (putting aside those novels that Nabokov had translated from Russian into English in the interval). In *Pale Fire,* the relation between the two worlds of Shade and Kinbote is antagonistic; both interpretations cannot be correct. They occupy the same space and time, but touch each other only tangentially. The curious echoes and pale reflections between them are all that provide Shade and Kinbote (though in very different ways) with the reassurance that life is anything more than an abscene and prolonged joke.

In *Ada,* on the other hand, the two worlds, Terra and Antiterra, are parallel and complementary, though occupying a different space and time; there is no difficulty in supposing both can be true at once. With so much reflection between them, the heavy hand of necessity can be seen not in the chance that there is no correspondence, but in the possibility that the correspondences may be too perfect. The emphasis thus shifts from *Pale Fire*'s "artistry of coincidence" to *Ada*'s "artistry of asymmetry." What Nabokov withholds from Kinbote in *Pale Fire* is the fundamental congruence that can confer the status of reality upon Kinbote's artistic dream. But what Nabokov withholds from his protagonist in *Ada* is perfect correspondence which, if granted, would result in sterility, if not in proof that at least one of the images did not exist. The distinction is the difference between the "tragic farce" (Carol Williams's phrase for a genre that she finds prototypical of Nabokov's fiction) of *Pale Fire* and the tragicomedy of *Ada.*

These differences between *Ada* and Nabokov's earlier work appear especially significant when we consider that it was while *Ada* was being written that Nabokov read Gardner's book, and possibly Gardner's later *Scientific American* article on the broken symmetries of physics. Already concerned with problems of artistic freedom in the midst of temporal necessity, with the need to redefine time as something other than linear sequence, and with the appeal of asymmetry, Nabokov found in Gardner a synthesis of these elements into a brilliantly simple thesis: amid the overwhelming symmetry of nature there exist slight asymmetries, and these asymmetries, collected together into antiworlds, imply that time can go backward. The result is a changed stance in Nabokov's work toward the relation between art and reality.

L. S. Dembo has pointed out that Nabokov's fiction before *Ada* had consistently embodied a tension between the protagonist's desire for a solipsistic world of his own making and the author's intervention to puncture that desire, revealing the inadequacy or even the insanity of the protagonist.[19] Such a technique presupposes that the author possesses a more secure ontological viewpoint than the narrator; otherwise, the tension could not exist. Nabokov's narratives are problematic not because reality does not exist, but because they are not reality.

Ada, however, represents a significant variation on the Nabokovian pattern of ontological security. In *Ada,* Nabokov seems quite self-consciously to have set himself the task of coming to terms with the new physics and, by implication, with the connection between art and the verifiable reality of scientific theory. Some readers, tempted by the usual Nabokovian pattern, have proposed that Antiterra is another solipsistic world of the narrator's creation.[20] But to accept this proposition is to simplify the text and ignore the kind of complexities that Nabokov is exploring. In *Ada,* the conflict is not between a world of illusion in which desires can be fulfilled and a real world that continually frustrates the artist's desire for control. Rather, it is the subtler tension inherent in a real world that seems partly to be amenable to the narrator's attempt to control it and partly to resist those patterns through its stubborn asymmetries.

The change is writ large in the novel's cosmology. Terra and Antiterra represent the two different kinds of worlds: one fully accessible to the protagonist but from our viewpoint unreal; the other real but shrouded in mystery for the protagonist. There is a mirror reversal here that keeps us from too easily equating Antiterra with the artificial world of art and Terra with reality, since for Van the terms of those equations are reversed. But it is significant that Van, far from ignoring Terra, spends his professional life attempting to communicate with it, just as it is significant that he is both scientist and artist. The scheme suggests that in *Ada* Nabokov is trying to connect his fictional, created world with the "real" world of scientific observation. If so, we may speculate that the enticement for Nabokov is the thesis that he found in Gardner: that time reversal is not merely an artistic dream, but has been verified by scientific observation.

[19]Dembo, pp. 3–18.

[20]See for example Bobbie Ann Mason, *Nabokov's Garden: A Guide to Ada* (Ann Arbor: Ardis, 1969). Mason's thesis is that Van, consumed by guilt over his incestuous relationship with his sister, invents Antiterra out of whole cloth; he really lives on earth the whole time.

This new conjunction between art and science has its own anxieties, however, for it must have raised questions about Nabokov's artistic control over his material. For Nabokov, science means "above all natural science," and hence a commitment to accurately describe a preexisting reality.[21] But when he speaks of his creative writing, he imagines himself as a "perfect dictator" who is "alone responsible" for the created world's "stability and truth."[22] To connect art and science, as Nabokov tries to do in *Ada,* would thus bring into conflict two opposing methodologies: one dedicated to the accurate observation of a preexisting reality; the other to the premise that art is an illusion under the complete control of its creator. The solution that Nabokov apparently arrived at was to appropriate the relativistic field model, but also to introduce into it the idiosyncratic variations that spring from his personal will as the creator of *Ada*'s universe. Thus, he asserts his own control at the same time that he avails himself of the legitimating power of the model to validate the reversal of time.

How Van distorts the Special and General Theories of Relativity is discussed at length by Strother Purdy in *The Hole in the Fabric.* Noting these distortions, Purdy remarks that if Nabokov himself understood the concepts, he kept it "well concealed."[23] Purdy points out, for example, that Antiterra is not really an antiworld in Gardner's terms, because it is not moving backward in time from our world, only parallel to it. Purdy concludes that Nabokov discards current scientific theories frivolously, flying "in the face . . . of relativistic time and the relation of space and time" in relativity theory, without offering any plausible solutions of his own.[24]

But in a sense Nabokov has offered a solution, or at least a compromise, to the larger question that Purdy's objection raises. At issue is the artist's control over his material, and the kind of restrictions he becomes subject to when he incorporates into his text a well-articulated scientific model. Purdy's position is extreme: he implies that Nabokov, to play fair, must faithfully reproduce the model in all its aspects. Purdy is

[21]The quotation is from "An Interview with Vladimir Nabokov," p. 33. For a fuller discussion of Nabokov's attitudes toward science, see Timothy F. Flower, "The Scientific Art of Nabokov's *Pale Fire,*" *Criticism,* 17 (Summer 1975), 224.

[22]"An Interview with Vladimir Nabokov," p. 25.

[23]Strother Purdy, *The Hole in the Fabric* (Pittsburgh: University of Pittsburgh Press, 1977), p. 127.

[24]Ibid., p. 131.

correct when he observes that Nabokov does not adhere to this requirement. Rather, Nabokov implicitly insists on his right to control his created universe by adhering to the model in some respects and rejecting or altering it in others. On a deeper level, however, Nabokov concedes Purdy's underlying premise that some restrictions on the artist are inevitable if he wishes to maintain contact with reality. Witness Nabokov's ironic stance toward his narrator: Van wants complete control, but Nabokov arranges matters so that this is equated with perfect reflection, and hence with the perfect mirror images that would trap the protagonist in a world of illusion. Though Van resists the knowledge, Nabokov knows that "a perfect likeness would rather suggest a specular, and hence speculatory, phenomenon" (p. 21), as one Antiterra scholar observes. The observation implies that if the world of art, cut loose from reality, is free to assert its primacy of being, it is also a retreat into a solipsistic, self-reflexive creation that is endlessly circular. Cut loose from reality, "all art [is] a game" (pp. 480–481). By admitting the asymmetries that modern field theory links with time reversal, Nabokov at once connects his fiction with scientific fact and implicitly avoids the circularity of self-reflexive art.

We are now in a position to meditate more deeply on what Antiterra means. As Purdy has pointed out, it is obviously not an antiworld in the strict scientific sense of being a world composed of antimatter and moving backward in time. Rather, it is a complementary and completing half of the arc of our world, an "antithetic" world that is a mirror reflection of our "thetic" reality, but with significant twists that keep it from exact mirror symmetry. It is precisely in those asymmetric twists that the circle is set free, so that the mutuality between the two worlds can be a creative rather than a destructive tension.[25] In that spirality, that lack of exact reflection, Nabokov can both exercise the control that

[25]Nancy Anne Zeller's "The Spiral of Time in *Ada*," in *A Book of Things about Nabokov* (Ann Arbor: Ardis, 1974), pp. 280–291, demonstrates how the chronology of Van's encounters with Ada can also be seen as taking place (with a little wrenching) along a spiral. Her thesis is that Van's last reunion with Ada is slightly off a perfect spiral, and that he therefore needs to turn back a year, as it were, by running down a spiral staircase, meeting Ada at last on one level below his present room. The symmetry and slight asymmetry Zeller notices are entirely consistent with my point here. Zeller seems not to take into account, however, that the helical spiral (a spiral of constant diameter) is only one possibility, and that spirals of increasing diameter are not only possible but indeed what Nabokov seems to have had in mind (the "still ampler arc" of the *Speak, Memory* passage).

he identifies with artistic creation and accommodate the demands of a scientific reality. Nabokov grants to science the limitations on artistic control; he gains back from it the assurance that time can indeed go backward.

The possibility that time could be reversed must have touched a deep chord in Nabokov. Even a casual reader of his autobiography, *Speak, Memory,* is aware of the crucial role that the sense of a lost past played in Nabokov's life. Exiled from his homeland, fated to see the entire way of life he knew there destroyed forever in the Revolution, Nabokov felt keenly the appeal of nostalgia. In *Ada* that past is recreated, history rewritten so that Nabokov's two homelands merge in "Amerussia." Van's attempt to blunt the point of time's arrow, denying its relentless forward flight by capturing, and then recreating, the ardors of Ardis through memory, is surely in part Nabokov's endeavor as well.

I have been suggesting that in *Ada,* the difficulty of reconciling art and reality lies not primarily in internal contradictions within the novel's created universe, as it does for many other Nabokov novels, but in Van's resistance to asymmetries. Antiterra and Terra both exist; the mirror reflections are not merely a function of Van's imagination. But the broken symmetry also insures that the search for a completely mutual reflection cannot be successful, and this in turn implies the failure of the narrator's attempts to control completely what can only partially be made to come under his shaping imagination. In this struggle, imagination is one important factor: it promises the possibility of free artistic creation. But in *Ada,* creation depends on both imagination *and* memory; and memory is tied to the accurate observation that Nabokov associates with science. If half of time's spiral is created through imagination, the other half is created through memory.

In the Appel interview, Nabokov commented that "imagination is a form of memory. . . . An image depends on the power of association, and association is supplied and prompted by memory. When we speak of a vivid individual recollection we are paying a compliment not to our capacity of retention but to Mnemosyne's mysterious foresight in having stored up this or that element which creative imagination may use when combining it with later recollections and inventions. In this sense, both memory and imagination are a negation of time."[26] As we have

[26]"An Interview with Vladimir Nabokov," p. 32.

seen, what Nabokov calls in his own life "Mnemosyne's mysterious foresight" is supplied to Van by his creator. Van need not invent his memories; he need only remember accurately to have the material at hand upon which his imagination can act. But memory also implies that the memoirist is constrained within the boundaries of accurate recall. Memory represents the "antithetic" constraint that exists in tension with the "thetic" freedom of the artistic imagination. In the broken symmetry of this conjunction, the spiral moves yet another turn, yielding a "still ampler arc" in Van's attempt to control time.

In *Ada* there are two different types of memory: true memory and false memory. Van makes the distinction most often by comparing his recollections to a camera still or movie. Both re-create events, but to Van, the mechanical recall provided by a camera or movie is a pale mockery of the sensual immediacy of his own memories. For example, Van sees Marina's sensibility as inferior to his and Ada's, and even to Demon's, because of her willingness to rely on mechanical reproduction rather than direct sensual recall of the event. Demon, faced with the women he had loved beyond all reason twenty years earlier, tries to "*possess* the reality of the fact by forcing it into the sensuous center" (p. 265). Marina, however, is content to remember their affair as a "stale melodrama" neatly filed in her "screen-corrupted" mind (p. 267). When Marina sees Van and Ada's hands together in 1888, she can't summon the memory it ought to trigger from 1884, the view of their twin hands gliding up the staircase rail in tandem ("though only *four* years had elapsed!" Van remarks in parenthetical exasperation). Marina is a "dummy in human disguise" because she lacks that "third sight (individual, magically detailed imagination)" which makes memory capable of vivid re-creation. Marina's flat recollections are distanced in her mind as if she were watching a movie of her own past—a movie she intends to edit and rearrange at her convenience.

By contrast, the sensual immediacy of Van's recollection is the very soul of his narrative. But he pays a price for this immediacy that Ada implicitly acknowledges when she says that "no point would there be, if we left out, for example, the little matter of prodigious individual awareness and young genius, which makes, in some cases, of this or that particular gasp an *unprecedented and unrepeatable event* in the continuum of life or at least a thematic anthemia of such events in a work of art, or a denouncer's article. The details that shine through or shade through . . . [are] all" (pp. 76–77; italics in original). The ambivalence in this passage (the "thematic anthemia," the "unrepeatable event" that Van desperately yearns to reproduce) "shades through" despite the celebration of vivid recall. For if the false memory of movie-like distancing leads to the purgatory of non-personhood, the true memory of vivid sensual recall can sometimes lead straight to hell. When Van recalls—in his usual blindingly clear fashion—the memory of Lucette's suicide "in a series of sixty-year-old actions which now I can grind into extinction only by working on a succession of words until the rhythm is right" (p. 521), the agony is unmistakable.

Van, unlike Marina, does not edit. He reports the sickening betrayals and falsities with the same fidelity of recall that he brings to the glorious moments. But he characteristically narrates shameful moments as though they were a movie script, perhaps unconsciously relegating them to the purgatory of false memory, which at such times is a relief from the hell of unrelenting immediacy. For example, in recollecting the episode where he and Ada try to seduce Lucette in bed between them (an episode he feels ashamed of afterward), he writes as though he were trying to force his sensuous recall into the impersonal angles of a camera panning across a ceiling mirror: "Thus seen from above . . . we have the large island of the bed illumined from our left (Lucette's right) by a lamp burning with a murmuring incandescence on the west-side bedtable. . . . Another trip from the port to the interior reveals the central girl's long white left thigh . . ." (pp. 443–444). Though the sensual immediacy keeps breaking through in that "long white left thigh," it only qualifies, it does not negate, Van's attempt to distance his shame by resorting to movie-like recall.

We may, if we wish to attribute a psychological motive to Van's association of movies and betrayal, trace it back to the summer of 1888, when Ada confesses her infidelity to him by paralleling herself with the unfaithful heroine of the movie script of *Les Maudits Enfants*. The entire chapter following this confession is written as a movie scenario: "If one dollied now to another group . . . one might take a medium shot of the young maestro's pregnant wife" (p. 210). The dreadful movie script dominates this chapter, seeming to dictate the actions of the characters out of the script as well as those in it. Poor Philip Rack, playing a part in life parallel to the wretched third lover who is included in the movie heroine's affections only because she pities him, confesses to Ada, "One feels . . . One feels . . . that one is merely playing a role and has forgotten the next speech" (p. 214). Van, included by proxy in the movie as another of the heroine's lovers, is also caught in the script when he strides away from pool-side at the same moment that the character who is his cinematic counterpart "leaves the pool-side patio" (p. 215).

Though Van's associations with movies are mostly negative, Ada can see the "sun side," for example when she explains why she *enjoys* acting within the confines of the script. "In 'real' life we are creatures of chance in an absolute void—unless we be artists ourselves, naturally; but in a good play I feel authored, I feel passed by the board of censors, I feel secure, with only a breathing blackness before me (instead of our Fourth-Wall Time)" (pp. 451–452). The constraints provided by the script thus provide Ada with a protection analogous to that Van finds in mirror reflection, that is, a predictable pattern that protects one from the anxiety of an indeterminate future cut loose from the bounds of the past.

As an artist, however, Van is unable to accept this solution because a script can never be totally under the artist's control. Van objects to making movies of books because the book "belonged solely to its creator and could not be spoken or enacted by a mime (as Ada insisted) without letting the deadly stab of another's mind destroy the artist in the very lair of his art" (p. 450). If Van is not in complete control of the emergent pattern, weaving it according to his preferred design, then he is unable to accept it as a valid extension of the past into the present.

Perhaps this is why Van, despite his dislike of movies in general, dotes on Ada's portrayal in *Don Juan's Last Fling*. By some "stroke of art, by some enchantment of chance," Van finds Ada's three brief scenes to be a "perfect compendium of her 1884 and 1888 and 1892 looks" (p. 520). Because the movie's portrayal exactly coincides with his own recollections, Van can accept the movie as an extension of his own memory. What he can't stand is a reification of memory that conflicts with his recall. Hence his outrage at Kim's pictures. Van finds the pictures not just embarrassing or inconvenient but a desecration, because they posit another version of the reality that Van recollects, a version not under his direct control. So he tells Ada, "This is the hearse of *ars*, a toilet roll of the Carte de Tendre! I'm sorry you showed it to me. That ape has vulgarized our mind-pictures" (p. 430).

Even though Ada welcomes constraints on action, it is emphasized that only the artist can create the comforting pattern that will contain without being so constrictive as to kill the life it holds within it. Ada thinks of the "novelistic" atmosphere of their room at the Three Swans as a "frame, as a form, something supporting and guarding life, otherwise unprovidenced on Desdemonia, where artists are the only gods" (p. 553). Van, more ambitious than Ada, wants to be the "god" who creates the constraints. The egocentricity of this position is intrinsic to Van's characterization. He may be faithful to his memory by including the painful episodes along with the joyous ones; but it is still *his* memory which he is reifying by writing his chronicle. Nabokov's vision, more complex than Van's, admits to constraints on the artistic re-creation through memory. "Unless we be artists ourselves," we will not have the privilege of preserving our memories; but only the great artist can hope to escape from the difficulties besetting the odyssey of memory, as it steers between the Scylla of exact symmetry and the Charybdis of asymmetries that nevertheless must be acknowledged.

In response to Kim's pictures, Van intends either to "horsewhip his eyes out or redeem our childhood by making a book out of it: *Ardis,* a family chronicle" (p. 430). We know he does both. But even so, the asymmetry of Kim's "wrenched" recollection cannot be escaped. Encapsulated within the larger recollections of *Ada,* which we understand as Van's attempt to reclaim the past on his own terms, is the smaller detail of Kim's pictures, modestly posing its own version of that recollection. To recollect faithfully and completely is to include the memory of counter-recollections which deny that the memoirist's reality is the only one. The very comprehensiveness of the mnemonic endeavor insures that the means for it unraveling are contained within it.

Van's success in controlling the memory-patterns is thus never total. His partial success hints that the larger endeavor in which he is engaged, the stopping of time, can also never be complete. For when "memory and imagination" cooperate in a "negation of time," there is one final qualification: human mortality. Only an immortal can remember forever. As Van admits, the inverse proposition is also true: "you lose your immortality when you lose your memory." He continues: "And if you land then on Terra Caelestis, with your pillow and chamberpot, you are made to room not with Shakespeare or even Longfellow, but with guitarists and cretins" (p. 622). Although in one sense death "catapults us altogether out of space, out of time," as

Martin Gardner puts it,[27] finally allowing us to achieve true timelessness, in another sense mortality represents time's final victory over the memorist, for when he dies his memories die. Van recognizes the paradox very early. "Space breaking away from time," what he strives to accomplish in his philosophical treatise, is fully achieved only "in the final tragic triumph of human cogitation: I am because I die" (p. 164). The only recourse is to do what Shakespeare and Longfellow did, and what the cretins do not: defeat the final amnesia by preserving the memories in lasting form, which is the real reason Van writes *Ada*. He hopes that he and Ada "will die *into* the finished book."

And in a sense they do. But Nabokov knows that recapturing life through art, like living memory, is transitory. Given enough time, the mnemonic enterprise to defeat Time must necessarily fail, since the full realization that Time can be overcome through the twin efforts of memory and imagination is limited to the duration of our own memory, fully achieved only as we see the completed pattern and the next moment already fading as we close the book. At the moment the narrative catches up with Van's present, Van finishes writing his book on Time, and his description of its emphasizes how fleeting is the moment when the meaning of the text is both fully immanent and sensually vivid: "My aim was to compose a kind of novella in the form of a treatise on the Texture of Time, an investigation of its veily substance, with illustrative metaphors gradually increasing, very gradually building up a logical love story, going from past to present, blossoming as a concrete story, and just as gradually reversing analogies and disintegrating again into bland abstraction" (p. 599).

It is fair to say this is Nabokov's strategy as well. For in *our* book, the past has been slowly accumulating meaning at the same time as it is moving toward Van's present, the present that we are aware of through narrative interjections but that we can glimpse only as momentary fragments distracting us from the nostalgic text. When the narrative reaches Van's present, Nabokov allows only a brief moment of synchronicity before the two time lines again begin to diverge, with us, the readers, going back into the reality of our presert on Terra, and Van and Ada fading into the indeterminacy of a future unknown to us. "Actually, we

[27]Gardner, *The Ambidextrous Universe: Mirror Asymmetry and Time-Reversed Worlds,* p. 272.

had passed through all that," Nabokov tells his readers after Van and Ada are finally reunited; "Our world *was,* in fact, mid-twentieth century. Terra convalesced after enduring the rack and the stake . . ." (p. 618). This transition back into our own time comes at the end of the penultimate chapter. As we begin the final chapter, Van and Ada grow progressively remote and fuzzy in outline, as they recede away from us back into "bland abstractions." The book ends with a bald plot summary of the book, wryly acknowledging its inability to recapture the reality of the lived experience. In another sense, of course, it is Nabokov's final attempt to do just that. For as we read the mock-blurb, we compare it with the story that now exists in *our* memory; and finding the memory infinitely richer than the summary, we are faced with the recognition of having achieved, through the art of Nabokov's words, a lived memory beside which his final verbal representation of it pales into parody. The beauty of Nabokov's art in *Ada* is the way he makes a virtue of necessity, achieving success through admitting partial failure. If the successes of *Ada* are never absolute—Van in his *Treatise on Time* admits he is "wounded by the Imposter" Space—the failures are never unmitigated by success. The only unqualified truth to emerge is the certainty that life and art are double, forever shade and sun intermingled.

The metaphysic of *Ada* may thus be defined as follows: what can be controlled is never completely real; what is real can never be completely controlled. The metaphysics is strikingly similar to Einstein's famous aphorism, "As far as the laws of mathematics refer to reality, they are not certain; and as far as they are certain, they do not refer to reality."[28] *Ada* is thus Nabokov's tribute to an idea intrinsic to the field concept, that reality can never be entirely captured in the abstractions of either art or science.

Yet there is a subtler implication of the field concept that Nabokov misses, or perhaps ignores. The desire for control is predicated on the assumption that it is possible to make an unambiguous separation between the one who controls and that which is controlled. It relies, in other words, on the Cartesian dichotomy, and hence is deeply bound up with the Newtonian world view. To accept that control can never be complete is to modify that world view, not to reject it; the revolutionary view would be to propose that the control is a chimera, and the

[28]Albert Einstein, *Ideas and Opinions,* ed. Carl Seelig (New York: Dell, 1973), p. 228.

desire for it is induced by the illusion that we are separate from the world and each other.

Nabokov's stance toward the field model is thus finally ambivalent. By imposing limitations on his narrator's attempt to control time through the twin efforts of imagination and memory, Nabokov voluntarily subjugates his art to some of the requirements imposed by a field model. The concession is reflected in the fact that in *Ada* Nabokov brings various versions of the two halves of a symmetric whole—Van and Ada, Terra and Antiterra, art and science—into tense and close relation. But relation is not unity; in admitting this much, Nabokov has escaped the more radical implications of the field concept. The image that lingers from *Ada* is not the universe made whole, but the atomistic universe of the Cartesian dichotomy compressed into two nearly perfect mirror images that, through their slight asymmetry, resist complete union. That tension is a measure both of the influence of the new physics on Nabokov's thought and of his ambivalence toward locating his fictional world within the cosmic web.

CHAPTER 5

SUBVERSION

Infinite Series and Transfinite Numbers in Borges's Fictions

> But can we not ask ourselves whether there do not exist, in the real universe, things which are non-algorithmizable, non-reducible, non-unifiable. . . . From that moment on, does the problem become, not to eject and repress uncertainty, randomness, disorder, antagonism outside its royal domain, but to seek dialogue with them?
>
> Edgar Morin, "Beyond Determinism"

NABOKOV AND BORGES are often compared, but their responses to the field concept are very different: whereas Nabokov is drawn to it because its asymmetries promise to rescue art from being merely a game, Borges is attracted to it because its discontinuities reveal that everything, including itself, is no more than a game. The two stances are associated with very different literary strategies. As we saw in Chapter 5, the impetus of *Ada* is to stop time; to create patterns whose parameters, once set, can encompass all future permutations; to make absolute and immortal the identity of the narrator, idiosyncrasies intact, by weaving his patterns of thought into the fabric of the created world.

Borges, by contrast, attempts to increase rather than use up the available permutations. Instead of hundreds of pages he writes five or six, characteristically including at least one open-ended catalogue capable of indefinite expansion. For Borges stasis is impossible because art is not an object to be framed, but a continuing process whose permutations are inexhaustible. In "Pierre Menard, Author of *Don Quixote*," for example, a changed context results in a completely different text. The *Don Quixote* of Pierre Menard, we are told, is a richer, subtler, and vastly *different* book from the *Don Quixote* of Cervantes, even though they use identical words. The assertion opens the way to an infinite number of *Don Quixotes*, all different once they are attributed to different writers and different historical contexts.[1] In "Tlön, Uqbar, Orbis Tertius," we learn that in Tlön literary critics attribute wildly different words to the same imaginary author, and then they explore the psychology of "this interesting *homme de lettres*" (*F*, p. 28). Different texts can thus give rise to a new author, just as imagining different authors can engender multiple texts from the same words. This view of art suggests that infinite sequence would be a natural—perhaps inevitable—metaphor for Borges to adopt.[2] What stasis is to Nabokov, sequence is to Borges.

Borges's strongest links to the field concept are through the infinite sets that Cantor introduced into mathematics in the latter nineteenth century. Before we turn to a fuller exploration of this connection, however, some history and definitions of the mathematical terms involved may be helpful. To understand Cantor set theory, it is necessary to understand the distinctions between sequences, series, and sets. (Although Borges uses these terms interchangeably, I will follow standard mathematical usage.) In mathematics, a sequence is a list of terms, for example, 1, 2, 3 . . . A series is the sum of terms associated with a given sequence, for example, 1 + 2 + 3 . . . A set is a collection of terms that can be combined in different ways; for example, the set [1, 2, 3] can be combined into six subsets: {1}, {2}, {3}, {1, 2}, {1, 3}, {2, 3}. An infinite series is the partial sums associated with a sequence that has an infinite number of terms, for example the sums we would get if we continued to add together the sequence of whole numbers 1, 2, 3 . . . to infinity. Infinite series can be divided into two groups: convergent series, in which the sums converge to a finite value, for example .3 + .03 + .003 + .0003 + .00003 . . . , which converges to 1/3; and divergent series, in which the sum of the terms cannot be expressed except as a mathematical formalism, because the sums do not "settle down" to a definable limit as the series progresses. The tendency of a series to diverge is not

[1]Jorge Luis Borges, "Pierre Menard, Author of *Don Quixote*," in *Ficciones*, ed. Anthony Kerrigan (New York: Grove, 1962) (hereafter *F*), pp. 45–56.

[2]In *El idioma de los argentinos* (Buenos Aires: n.p., 1928), Borges argues that the richness of a language cannot be equated with the number of words it contains, since "el solo idioma infinito—el de las matemáticas—se basta con una docena de signos para no dejarse distanciar por número alguno" (p. 170).

always intuitively obvious. For example, the seemingly innocent progression 1 − 1/2 − 1/3 − 1/4 . . . is a famous instance of a divergent series. If the signs are alternately positive and negative (1 − 1/2 + 1/3 − 1/4 . . .), however, the series converges. A convergent series thus consists of the partial sums of an infinite sequence that nevertheless add up to a finite number, while a divergent series adds up to no definable sum at all.[3]

One of the earliest challenges to a deterministic universe came from infinite series; Zeno's famous paradox of Achilles and the tortoise uses the infinite series 1 + 1/10 + 1/100 + 1/1000 . . . to achieve its effect. As mathematicians soon discovered, however, the threat posed to determinism by infinite series could be defused by outlawing from mathematics actually infinite (as opposed to potentially infinite) sets of numbers. Consider for example the sequence 1, 2, 3 . . . As long as we consider the sequence to consist only of those numbers we have named so far, the infinite progression implied by the ellipsis is only potentially present, looming in the future but never actually reached. If, however, we consider the entire sequence as a pre-existing entity, the sequence is not the progression but the complete set of whole numbers, which are infinite in number. This set is then actually rather than potentially infinite, since there are an infinite number of terms present in it.

It had been traditional since Aristotle to consider actually infinite sets as illegitimate entities, and to think of "infinity" as a single, unimaginably large number. All that changed, however, with Georg Cantor. Cantor argued that actually infinite sets were legitimate mathematical entities; he also maintained that it was possible to determine the relative magnitude of infinite sets, and consequently to prove that some were larger than others. Since Cantor believed that infinite sets obeyed their own peculiar arithmetic, he referred to them not as infinities but as "transfinite numbers." Cantor chose the aleph to represent these transfinite numbers, designating his smallest transfinite number, the set of whole numbers, as $\aleph_0$; the next largest infinite set, the set of real numbers, as $\aleph_1$; the next largest as $\aleph_2$, and so forth. Cantor then proved theorems about the relationship between these transfinite numbers that became the basis for Cantor set theory. We shall return to Cantor set theory when we discuss Borges's appropriation of it in "The Aleph."

[3]Kennan T. Smith, in his *Primer of Modern Analysis,* clarifies the distinction between sequences and series on pp. 90–92.

Cantor set theory played a crucial role in the developments leading to the discovery of Gödel's theorem, and hence to those aspects of the field concept that emphasize self-referentiality. Although it is not correct, strictly speaking, to talk about a "field" in connection with Cantor ("field" has a different meaning in mathematical theories of groups), Cantor's unique contribution was to consider all the elements of an infinite set to be present (as he said) "at once," and thus to posit the number system as a pre-existing, interrelated totality. In making this shift, Cantor introduced into mathematics ambiguities and indeterminacies that we have seen to be characteristic of the field concept. Cantor himself realized this; he discovered that his set theory led unavoidably to paradoxes of self-reference.

Early in his work Cantor had demonstrated that for any infinite set, there is another one that is larger. Consider for example the set {1, 2, 3}. This set has three elements, and as we have seen, we can form from it six different combinations of these elements, or six subsets. We can intuitively appreciate, then, that the number of all possible subsets of a set will be larger than the original set. Since any set can be broken into its subsets, there will thus always be a number larger than any of the transfinite numbers. The problem arose in 1895 when, as Morris Kline explains, "Cantor thought to consider the set of all sets."[4] Since this set includes all possible sets, its number should be the largest that can exist. However, Cantor had shown that the set of all subsets of any given infinite set must be larger than the original set itself. Hence if the set of all sets is broken into its subsets, the number of these subsets must be larger than the original set, which implies that there must be a number larger than the largest number. The problem arises because the set of all sets contains itself as a member, and so refers self-referentially to itself. Cantor was never able satisfactorily to resolve the paradox posed by the set of all sets. He finally said merely that it was necessary to distinguish between consistent and inconsistent sets, and that self-referential sets were inconsistent.[5]

By asserting that actually infinite sets were legitimate mathematical entities and then discovering that they led to apparently irresolvable paradoxes, Cantor paved the way for other mathematicians to recognize related paradoxes that were not merely in set theory, but deeply embed-

[4]Morris Kline, *Mathematics: The Loss of Certainty* (New York: Oxford University Press, 1980), p. 203.
[5]Ibid., p. 202.

ded in the foundations of mathematics. In an attempt to eliminate these paradoxes, Bertrand Russell suggested that statements referencing themselves should be outlawed from mathematics. Unfortunately, self-referencing statements (or, in Russell's terminology, "impredicative definitions") had been used in classical mathematics in important definitions, and it was not clear how they could be banished without abandoning some very useful and widely held theorems. The controversy became part of what Kline calls the "loss of certainty" in mathematics, resulting in the realization that mathematics is not the absolute truth it was once supposed to be, but is riddled with unprovable assumptions and irresolvable paradoxes. The controversy over Cantor set theory led directly to the attempt to axiomatize number theory, and this, as we saw in Chapter 2, led in turn to Gödel's proof that complete axiomatization is impossible. Cantor was thus inadvertently a seminal figure in the chain of events culminating in the realization that the field concept implies inescapable limits on what can be proved by logical analysis.

Borges's discussion of Cantor set theory in *The History of Eternity* shows him to be well in command of these basic concepts. Borges's familiarity with the history and import of Cantor set theory is further indicated by his review of Kasner and Newman's popularized account in *Mathematics and the Imagination,* shortly after it appeared in 1940.[6] As his review makes clear, Borges not only understood Cantor's essential methodology, but also appreciated that it led directly to the discovery of paradoxes of self-referentiality. In his review Borges mentions volumes in his library that he has "made hazy with manuscript notes," and predicts that *Mathematics and the Imagination* will take its place among those well-read few.[7] Borges then goes on to list the "spells of mathematics" which are for him the most powerful; Strange Loops are the essence of this list. "There are almost innumerable versions of this method [of constructing self-referential paradoxes] that don't vary except in the protagonists or the fable," Borges notes. He then proceeds to list some of his favorites, including Bertrand Russell's discussion of "the set of all sets that don't [sic] include themselves" (*D,* p. 166).

Borges's review suggests that he was drawn to Cantor's work because

[6]Edward Kasner and James Newman, *Mathematics and the Imagination* (New York: Simon & Schuster, 1940).

[7]Jorge Luis Borges, *Discusión* (hereafter *D*), vol. 6 of *Obras Completas* (Buenos Aires: Emecé Editores, 1957), pp. 165–166.

he saw in it possibilities for creating new kinds of Strange Loops. As we saw in Chapter 2, it is when everything is connected to everything else through the mediating field that self-reference becomes a problem. It surfaces in its strongest form in the logical paradoxes of Cantor set theory. Borges is the first writer in this study who consistently wants to exploit rather than suppress these inconsistencies, because he hopes to use them to reveal the essential fictionality of the model. His intent is thus subversive.

His strategy is seduction, for he progresses to this revelation by several seemingly innocuous steps. The first step in his strategy is to transform a continuity into a succession of points, and to suggest that these points form a sequence; there follows the insinuation that the sequence progresses beyond the expected terminus to stretch into infinity; then the sequence is folded back on itself, so that closure becomes impossible because of the endless, paradoxical circling of a self-referential system. This complex strategy (which may not appear in its entirety in any given story) has the effect of dissolving the relation of the story to reality, so that the story becomes an autonomous object existing independently of any reality. The final step is to suggest that our world, like the fiction, is a self-contained entity whose connection with reality is problematic or nonexistent.

Borges's best-known stories, particularly those anthologized in *Ficciones* and *The Aleph and Other Stories,* show the strategy in prototypical form. Although these stories have often been written about, the role of infinite sequences in them has not been generally recognized. The omission is the more surprising because Borges's work is so repetitive: the same themes, ideas, and paradoxes keep recurring.[8] Borges himself is perfectly aware of this sparseness. In "Profession of Literary Faith," he writes, "I have already overcome my poverty; I have recognized, among thousands, the nine or ten words that accord with my heart."[9] But one of those "nine or ten words"—the concept of series—has yet to be fully

[8]The repetitiveness of Borges's work has been noticed by many critics, among them James E. Irby in his discussion of the mirror, the labyrinth, and the book as central symbols in Borges, *The Structure of the Stories of Jorge Luis Borges* (Ann Arbor: University of Michigan Microfilms, 1962), pp. 270–285. How easily Borges's work may be classified in terms of its central metaphors may be seen in the way Ana María Barrenechea organizes her material in *La expresión de la irrealidad en la obra de Jorge Luis Borges* (Buenos Aires: Ed. Paidos, 1967), trans. as *Borges the Labyrinth Maker,* trans. Robert Lima (New York: New York University Press, 1965).

[9]From *El tamaño de mi esperanza,* quoted in Ronald Christ, *The Narrow Act: Borges' Art of Illusion* (New York: New York University Press, 1969), p. 11.

recognized.[10] We will therefore consider a few of these stories to see how the general strategy works, before turning to its more specific application in "The Aleph."

The well-known "Tlön, Uqbar, Orbis Tertius" provides an example of the first steps in Borges's subversive strategy. The story opens in the familiar setting of casual after-dinner conversation, with Bioy Casares recalling that "one of the heresiarchs of Uqbar had stated that mirrors and copulation are abominable, since they both multiply the numbers of men" (*F*, p. 23). Soon this predicted series of specular worlds begins to appear in earnest. The source of the aphorism, we learn, is an elusive article in *The Anglo-American Cyclopedia* on the apparently nonexistent country of Uqbar. In the article, reference is made to Tlön, an imaginary region that is the subject of Uqbar's literature. Tlön, a fantastic region written about in a country which already cannot be located on any map, thus exists at several removes from reality, having its being for the moment only as a passing reference in a single variant copy of a pirated encyclopedia. With the discovery a few years later of the eleventh volume of *A First Encyclopedia of Tlön,* however, Tlön begins to emerge from textuality into actuality. In the end, Borges tells us, the world *becomes* Tlön. As we witness this transformation, we can see Borges's strategy evolving from the first tenuous suggestion of a sequence to the infinite progression of a self-referential Strange Loop.

Series, particularly nuances in number systems, are crucial to this transformation. A footnote intimates that in Tlön, base twelve rather than base ten is used for counting.[11] Gradually, we are led to suspect that base twelve is also the operative counting system within the story, which implies that as the world becomes Tlön, the text is also becoming a Tlönist document. Borges achieves this modulation by a subtle emphasis on the number eleven—a number that encourages us to proceed beyond the expected terminus of the decimal system. The volume of Tlön's *First Encyclopedia* that Borges first uncovers is the eleventh; heresies of the eleventh century are documented; *hrönir* of the eleventh

[10]Barrenechea, for example, despite her thorough and perceptive analysis of Borges's major motifs, does not even list "series" in the index.

[11]The footnote glosses the phrase "eleventh century heresiarch of Tlön" by explaining that in the duodecimal system, a century is composed of one hundred and forty-four years. The number is the square of twelve or twelve twelves, analogous to the hundred years or ten tens that comprise a century in the decimal system. The footnote thus indirectly establishes that in Tlön, the duodecimal rather than the decimal system is used for counting.

degree are emphasized as having "a purity of form which the originals do not possess" (*F,* p. 30).

The transition number, eleven, has a dual function. It leads us beyond the decimal system, yet falls short of the expected terminus of the duodecimal system. The withholding of the final term is important, for repeatedly a terminus is suggested, only to be transformed into an indefinitely continuing sequence. The eleventh volume of the *First Encyclopedia,* for example, though at first it is the only volume in existence, refers to "both subsequent and preceding volumes"; its "apparent contradictions" (*F,* p. 22) provide the basis for proving that other volumes exist, volumes which will in turn be superseded by the yet more numerous volumes of the succeeding edition. *Hrönir* of the eleventh degree are purer than the originals, suggesting a point of termination; but *hrönir* of the twelfth degree "deteriorate in quality," so the process "is a recurrent one" (*F,* p. 30). Even within the duodecimal system, the final term never quite arrives.

As single terms stretch into sequences, and as sequences modulate into nonterminating progressions, the locus of reality becomes correspondingly indeterminate. In the idealist metaphysics of Tlön, there is no reality independent of one's thoughts of reality; it is sufficient to search for a thing to bring it into existence. In Tlön the *ur* is a "thing produced by suggestion, an object brought into being by hope" (*F,* p. 30). There is the suggestion that Tlön itself is an *ur*-object, a reality brought into being by believing it exists. As scholars, alerted to the mystery of Tlön, begin searching for evidence of its existence, more and more evidence in fact appears, from the missing volumes of *A First Encyclopedia* to Tlönist artifacts. The implied progression suggests that the Terran scholars created Tlön by searching for it.

Meanwhile, the theorists of Tlön are reconstructing (or constructing?) our world through the parable of the nine coins, in which they try to imagine the continuous existence of matter through time. As eleven is the penultimate term of the duodecimal system, nine is the penultimate term of the decimal system. Our world conceives of Tlön in the eleventh volume; Tlön conceives of us in nine coins. Through the conjunction of the penultimate terms of the two number systems, each world seems to be evoking the other as the final, inevitable, and yet almost unimaginable last term of the alien number system it entertains.

But this terminus, like others in the story, is illusory. If our world can become Tlön, and Tlön can become our world, the exchange can take place again—and again. At some future date, when we have become Tlön, our philosophers will propose the paradox of the nine coins. Meanwhile, in the world that was once Tlön and is now a materialist society, a variant text appears alluding to a mysterious region where any other philosophy except idealism is inconceivable. The sequence thus posited implies that the two worlds will become each other in turn, each calling the opposite world into being through the penultimate terms of a sequence that never ends. Text metamorphoses into context, context into text, text into context, in a Strange Loop that makes the distinction between "fiction" and "reality" an undecidable question.

An even more bizarre sequence, based on another progression implicit in the text, is possible. On an occasion when the narrator is conversing with Herbert Ashe, whom we later learn is one of the "demiurges" responsible for creating Tlön, the discussion turns to the "duodecimal number system." Why Ashe, one of the creators of the duodecimal-based Tlön, should want to construct tables converting decimals to duodecimals is quite clear in retrospect to us and to the narrator, who laments that "nothing more was said—God forgive me—of duodecimal functions" (*F,* p. 21).

Ashe's conversion tables, however, do not end there. The narrator recalls Ashe's comment that "as a matter of fact, he was transcribing some duodecimal tables . . . into sexagesimals (in which sixty is written 10), adding that this work has been commissioned . . ." (*F,* p. 21). Apparently the plan of the secret society to which Ashe belongs is that the incredible world of Tlön, based on a duodecimal system, is to be succeeded by the unimaginable world of Orbis Tertius, based on the sexagesimal number system. Tlön was brought into being through the discovery of *A First Encyclopedia of Tlön;* the adjective "first" implies a sequence, a second term. Are the forty volumes of the first edition to be followed, as the narrator suspects, by the hundred volumes of *A Second Encyclopedia,* "another work, more detailed, and this time written, not in English, but in some one of the languages of Tlön" (*F,* p. 32), which in turn will call the third world of Orbis Tertius into being? The progression implies a sequence of worlds, each calling its successor into being in an increasingly unimaginable sequence that has no end. Infinite sequences, by combining known terms with an unattainable end point, exhibit the paradoxical qualities of boundedness and infinite regress. In Borges's story, as each term comes into view, another term looms behind it, so that regardless of how many terms are accepted, with whatever resignation, statis is never achieved.

The sequences in this story, however, do not only contain inherent paradoxes; they also harbor an inescapable contradiction.[12] The contradiction arises in the following way. We are given to understand that Tlön is a world of "extreme idealism." In Tlön's idealist metaphysics, each state of being is conceived of as separate and unconnected with preceding and succeeding states. Tlönist philosophy would thus deny the possibility of a sequence, since the terms of a sequence by definition form a connected progression. So in creating Tlön as one of the worlds that exist in a connected sequence, Borges has posited a sequence which contains within it a term, Tlön, that denies the possiblity of sequence. The internal contradiction destroys whatever feeling of certainty we may derive from inferring that the sequence will proceed in a stable progression. Stasis—even the stasis of constant metamorphosis through the terms of a sequence—is undercut to leave us in a fluid state where nothing is certain. "Now, in all memories," Borges writes, "a fictitious past occupies the place of any other. We know nothing about it with any certainty, not even that it is false" (*F,* p. 34).

Jaime Alazraki has suggested that the obvious parody behind the forty volumes of *A First Encyclopedia of Tlön* is the *Encyclopedia Britannica.* He infers from this that Tlön, created by a secret society of geographers, chemists, artists, and algebraists, is a thinly veiled metaphor for our own world, which is also a social construct created by a society of chemists, geographers, artists, and algebraists, and described in an encyclopedia.[13] Might we infer, then, that Borges is re-creating in this story an elusiveness that he sees as characteristic of reality? This hypothesis implies that we have some notion of what reality is—that is, that it has the quality of regressing before us. But even this modest hypothesis Borges sabotages in the following often-cited passage.

[12]Frances Wyers Weber, in "Borges' Stories: Fiction and Philosophy," *Hispanic Review,* 36 (1968), 124–141, makes the point generally about Borges's fictions that they are "self-reversing tales." What happens with series is a special example of this general paradigm.

[13]Jaime Alazraki, "Tlön y Asterión: Metáforas Epistemológicas," in *Jorge Luis Borges,* ed. Jaime Alazraki, No. 88 in *El Escritor y la Crítica* (Madrid: Taurus Ediciones, S.A., 1976), pp. 192ff.

> Almost immediately, reality gave ground on more than one point. The truth is that it hankered to give ground. Ten years ago, any symmetrical system whatsoever which gave the appearance of order—dialectical materialism, anti-Semitism, Nazism—was enough to fascinate men. Why not fall under the spell of Tlön and submit to the minute and vast evidence of an ordered planet? Useless to reply that reality, too, is ordered. It may be so, but in accordance with divine laws—I translate: inhuman laws—which we will never completely perceive. Tlön may be a labyrinth, but it is a labyrinth plotted by men, a labyrinth destined to be deciphered by men. (*F*, p. 34).

According to the narrator, we can know nothing about reality. All we can be sure of is that if we may decipher its principles, then it is not reality but a labyrinth, that is to say, an artifact. Even the assumption that the artifact reproduces the order characteristic of our experience of reality is tenuous, since according to the idealist philosophy of Tlön that experience itself is a creation of our own minds. It is therefore not possible to assume that the fiction is verisimilar.

It is also, oddly enough, not altogether possible to assume that it is *not* verisimilar. This further complication arises from the internal contradictions within the story. That the fictional world is inconsistent means that we cannot be sure if even the artifact is ordered, which in a curious way makes it again verisimilar to reality. If it is impossible to arrive at the final truth within the fiction, it may be like reality after all, its very inaccessibility rendering it mimetic. The reasoning is, of course, circular.

This leads us to the next step in Borges's strategy: after he insinuates a sequence, he subverts it by making it a circle or by making the sequence consist of a single term that repeats itself endlessly. In this turning of a sequence back on itself, Strange Loops can appear that render reality itself an undecidable proposition. The elaborate, circuitous turnings in "The Approach to al-Mu'tasim" illustrate the technique.

The form of this story is a review of a labyrinthine novel about the fabled al-Mu'tasim. At first the reviewer assumes that the solution to the labyrinth can be found when the protagonist finds the man for whom he searches. Almost unnoticed is an antecedent to this proposition: that in suggesting the search takes place through progressive revelations, the reviewer has transformed the narrative continuity of the novel into a sequence. The strategy becomes explicit when he suggests, "A mathematical analogy may be helpful here. Bahadur's populous novel is an ascendant progression whose last term is the foreshadowed 'man called al-Mu'tasim.' "[14] When the novel's hero draws the curtain from al-Mu'tasim at the novel's end, it seems that the terminus has been achieved.

That termination is soon undercut, however, by the reviewer's revelation of a 1934 version of the novel which "declines into allegory," an allegory in which al-Mu'tasim is a God who "is also in search of Someone, and that Someone of Someone above him . . . and so on to the End (or rather, Endlessness) of Time, or perhaps cyclically" (*A*, p. 50). As the apparent Source recedes into an unattainable end point, undercurrents of circularity further complicate the idea of the pilgrimage as a linear progression to a finite terminus. The hero's journey traces a circle, beginning in Bombay and ending there. It begins with a riot between the Muslims and the Hindus, "God the Indivisible against the many Gods" (*A*, p. 46). But this difference collapses into identity when we learn that different men see God differently, according to the biases of their religions; therefore the One is also the Many, appearing to the "many varieties of mankind" in the multiform guises demanded by their differing expectations. Because the One and the Many are indistinguishable, the religious dispute responsible for beginning the pilgrimage is based on an illusion; but this realization comes about only by virtue of the pilgrimage. The reviewer drily comments that "the idea is not greatly exciting" (*A*, p. 50), without noting that it makes of the pilgrimage an intellectual as well as a physical circle.

Another idea excites the reviewer more. In the twentieth chapter, certain phrases attributed to al-Mu'tasim appear to be "the mere heightening of others spoken by the hero; this and other hidden analogies may stand for the identity of the Seeker with the Sought" (*A*, p. 50). Again difference collapses into identity as the reviewer remarks, with some exasperation, that the novel's supposed author, Mir Bahdur Ali, "cannot refrain from the grossest temptation of art—that of being a genius" (*A*, p. 51).

What the reviewer chooses not to recognize is related circularities in

[14]Jorge Luis Borges, "The Approach to Al-Mu'tasim," *The Aleph and Other Stories, 1933–1969,* ed. and trans. Norman Thomas di Giovanni in collaboration with Borges (New York: E. P. Dutton, 1970) (hereafter *A*), p. 49.

his own essay, beginning with the fact that he is creating the novel by describing it. But the illusion of difference is important, for it allows the reviewer also to be a seeker, searching for the novel's meaning. His search at first seems to be moving away from the novel in a linear progression. He begins with a plot summary of the novel; then he gives various interpretations of the difference between its two versions; finally he adds a concluding section in which he adduces various literary analogues, from Kipling to *The Faerie Queene,* putting forth "the Jerusalem Kabbalist Isaac Luria" as his own choice for the novel's source. The narrator's attempt to move closer to the original 1932 version by first examining the 1934 version (the only one he has seen) and then adducing increasingly remote literary predecessors seems perversely to be carrying him in the wrong direction. But in the most remote section of all, a footnote on a possible literary analogue in the parable of the Simurgh, the narrator's pilgrimage is also revealed as a circle. For the footnotes confirms that the parable of the Simurgh is really another version of the novel. The thirty birds seeking the Simurgh realize that, by virtue of their arduous journey and consequent purification, they have *become* the Simurgh. The Seeker is again revealed as identical with the Sought, the journey with its end. The footnote ends with Plotinus's "divine extension" of the principle of identity: "'All things in the intelligible heavens are in all places. Any one thing is all other things. The sun is all the stars, and each star is all the other stars and the sun'" (*A*, p. 52). The identity of the footnote, an appendage to the addendum to the review of the text, with the text's central meaning is another way of converting the linear into the circular, and of confirming the principle of identity laid down by Plotinus.

The narrator as seeker after the novel is identical with that which he seeks in another sense, for the novel does not exist outside his essay. His exasperated admiration for the novel's complexities, his stance as a critic examining the novel objectively, create an illusion of difference that collapses in the final realization of identity as the circular, circuitous journey of the essay brings him back, through the footnote, to the core of the novel, which is also his review. The transformation of apparent linearities into circularities implies that the sequence closes back on itself, so that differences collapse into similarities and eventually into identity. This is, of course, precisely the kind of circling back that is characteristic of self-referential systems. The sense of moving along, of coming closer to an end point, is revealed not exactly as an illusion but as a half-truth as the seeker merges with the sought, the periphery with the center, the journey with its end. The effect is to create a self-referential field from elements that we initially take to be separate and distinct.

Where Borges's fiction differs from scientific models of the field concept, however, is in using the concept to suggest that everything, including reality, is a fiction. Scientific models, by contrast, are useful only because they are presumed in some way to reflect reality. The difference leads to very different views of Strange Loops in science and in Borges. For Cantor, for example, Strange Loops were an embarrassment, but for Borges they are an embarrassment of riches because they allow him to draw into question the assumption that there is a "reality" to reflect. This predilection of Borges's is clear in "The Library of Babel," a fable in which the library and the world become synonymous. In this story designed to draw our attention to the artificiality of both the word and the world, we see the final step in Borges's seductive strategy, the inclusion of the reader himself in the circle of the fiction's Strange Loop. The narrator of this story, who refers to himself as a Librarian, is convinced that the intensely symmetrical order of the Library has to conceal meaning somewhere among its symmetries. But he (and we) slowly realize that order is not the same as meaning. The dilemma is extended to cosmic proportions when the narrator comes to what he calls the "capital fact" of his history. Clearly the Library must be infinite, the narrator argues; but the number of books derivable from all possible combinations of letters, though very large, is still finite. The narrator suggests that the answer to the dilemma lies in realizing that finite terms can become an infinite series if they are continued indefinitely. After exhausting the permutations provided by the twenty-five orthographic symbols the Library simply repeats itself, thereby becoming "limitless and periodic." The narrator speculates that the larger sequence of repeated terms would then possess an order, the order of repetition. "If an eternal voyager were to traverse [the Library] in any direction," the narrator concludes, "he would find, after many centuries, that the same volumes are repeated in the same disorder (which, repeated, would constitute an order: Order itself)" (*F*, pp. 87–88). "My solitude rejoices in this elegant hope," he ends.

But the narrator's "solution" is of course an answer only in a very narrow sense. While it suggests a way to transform randomness into ordered sequence, it contains no hint of how that sequence may be rendered intelligible or meaningful. Even if order is assured, sense is not; the problem of what the Library or its books mean remains unanswered. The narrator's solution is significant not because it is satisfying, but because it is inevitable. It reveals how desperate is the human need to find meaning somewhere, even if only in the repeated order of meaningless sequence. "Let me be outraged and annihilated," the narrator prays, "but may Thy enormous Library be justified, for one instant, in one being" (*F,* p. 86).

Almost as inevitable is the appearance of a Strange Loop; this time it extends to encompass the reader. The narrative we read is contained in a book, the book we hold in our hands. But if, as the narrator claims, the Library contains all possible books, an identical text is also somewhere among the Library's volumes. Which then are we reading, the narrator's history or the Library's book? The narrator specifically contrasts his "fallible hand" scribbling his story on the end leaves of a book with "the organic letters inside: exact, delicate, intensely black, inimitably symmetric" (*F,* p. 81). Logic demands that we conclude the present text in hand (which of course is printed) to be the Library's book. What we have is not the narrator's handwritten text but a mirror of it, or perhaps one of the "several hundreds of thousands of imperfect facsimiles" (*F,* p. 85).

That the account is inaccurate in indeterminable ways is one implication; even more important is the implication that we are reading the Library's book. This in turn implies that we, like the narrator, are within the Library examining one of its volumes, which means that we, no less than the narrator, are contained within one of the books we peruse. The narrator's remark that "the certainty that everything has been already written nullifies or makes phantoms of us all" (*F,* p. 87) takes on a disturbing new meaning as we realize that we too must be within one of the Library's books. "Why does it disquiet us to know that Don Quixote is a reader of the *Quixote,* and Hamlet is a spectator of *Hamlet?*" Borges asks in "Partial Enchantments of the *Quixote.*" "I believe I have found the answer: those inversions suggest that if the characters in a story can be readers or spectators, then we, their readers and spectators, can be fictitious."[15] With this realization comes the

[15]Jorge Luis Borges, "Partial Enchantments of the *Quixote,*" in *Other Inquisitions, 1937–*

corollary awareness that the Strange Loop now encompasses us within its circumference.

Sequences thus play an important role in allowing Borges to develop the idea that both his fictions and the world are self-referential systems. They are no less important in providing him with a model for the structure of his stories. The possibilities are implicit in Zeno's Second Paradox, the parable of the tortoise who has a head start and goes one length before the hare begins. But after the hare has gone one length, the tortoise has gone one and one-tenth lengths, so the tortoise is still ahead. The paradox lies in the fact that no matter how far the hare goes, the tortoise is always one-tenth the distance ahead. The argument owes its force, Borges points out in "Avatars of the Tortoise," to the series 1 + 1/10 + 1/100 + 1/1,000 + 1/10,000 . . . (*OI*, p. 110). The method suggests that infinity can be made to stretch between any two points simply by converting distance from a continuum to an infinite series of decreasing magnitude.

We have seen how Borges uses this idea in developing the plots of his stories; he also uses it in constructing the paradoxes that reveal the fictionality of the model. Continuity, particularly continuity of space and time, implies that moving from point A to point B requires traversing the intermediate distance. Borges prefers, on the contrary, to break continuities into sequences so that point A suddenly turns into point B (or point Z). Because the transition between terms is discontinuous, it is also possible for Borges to suggest that space intervenes between terms—and into this space he can insert a sub-series of infinite length. He thus gains two advantages from sequences that he could not derive from continuities: the ability to effect abrupt transitions between discontinuous terms (for example, by positing opposites as terms of the same sequence so that progression along the sequence suddenly transforms one thing into its opposite); and the ability to suggest that infinity can stretch between any two terms of the sequence, thus rendering any real progression an impossibility.

In addition to its thematic uses, the inherent discontinuity of a sequence provides a model for the narrative structure of the stories. The "middle distance" that narrative fiction usually extends is excised in Borges through ellipses and sudden breaks that transform a chronologically continuous story into a series of disconnected narrative points.

1952, trans. Ruth L. C. Sims (Austin and London: University of Texas Press, 1964) (hereafter *OI*), p. 46.

With the collapse of the "middle distance," the length we traverse is defined less by the duration of the narrative than by the implied distance contained between the ellipses (which can be infinite). It is as if the linear spatiality of narrative fiction has been transformed into the repeated downward plunges one might endure in attempting literally to traverse a sequence. In this metaphorical, Borgesian view of sequence, the spaces between terms are as important as the terms themselves. The discontinuities facilitate the abrupt leap of thought necessary for the essential paradoxes to emerge. They also create gaps that make it impossible even to estimate where the bottom might be. The sequence, implicit chasms yawning between its apparently adjacent terms, thus plays a metaphoric as well as conceptual role in the fictions, serving as both sign and symbol in Borges's fictional mode.

So far I have been speaking of a sequence as an infinite progression of discontinuous terms. We are ready now, however, to consider the Cantorian view of an infinite series as a single entity, a pre-existing whole defined by the entire sum of terms. As we have seen, this view of series allows infinity to be encapsulated, as it were, within the bounds defined by the initial and final terms. Borges, aware of Cantor's work, sees in transfinite set theory new possibilities for representing infinity within the finite bounds of his short stories—and also new threats to the "wildness" that he loves in infinity. Never merely a reporter of ideas, Borges subjects Cantor's theory to what Harold Bloom would call a "strong reading," deforming it in ways that reveal what he hopes to gain by infiltrating his fictions with infinite series and sequences.

Cantor was able to answer the question "How large is infinity?" by placing infinite sets in one-to-one correspondence. "The operation of counting is nothing more for [Cantor] than that of comparing two series," Borges comments in his discussion of Cantor set theory in "The Doctrine of Cycles."[16] The idea is to match up each element of set A with an element from set B; if it can be shown that the matching is complete, with no elements left over in either set, then the two sets can be said to be equal in size, even though the number of elements in each is infinite and therefore not countable. As Borges points out, we can show that the number of odd numbers is equal to the number of even numbers by using this one-to-one correspondence method (*BR,* p. 67):

[16]Jorge Luis Borges, "The Doctrine of Cycles," in *Borges: A Reader,* ed. Emir Monegal and Alistair Reid (New York: E. P. Dutton, 1981) (hereafter *BR),* pp. 66–67.

1 corresponds to 2
3 corresponds to 4
5 corresponds to 6, etc.

Though Borges does not say so, it is also possible to show that the number of odd numbers alone is equal to the total number of both odd and even numbers by a similar procedure:

1 corresponds to 1
3 corresponds to 2
5 corresponds to 3
7 corresponds to 4, etc.

Using this method, Cantor arrived at the paradoxical result that a subset (or partial grouping) of an infinite set is as large as the complete set itself. In fact, Cantor defined an infinite set as a set that can be put into one-to-one correspondence with one of its proper subsets. Borges makes the correct, but to a non-mathematician surprising, observation that "there are as many multiples of 3018 as there are numbers [that exist]—without excluding from these 3018 and its multiples" (*BR,* p. 67). The proof follows the earlier method:

1 corresponds to 3,018
2 corresponds to 6,036
3 corresponds to 9,054
4 corresponds to 12,072, etc.

What fascinates Borges about these results from Cantor set theory is the idea that "in these elevated latitudes of enumeration, the part is no less copious than the whole: the exact quantity of points there are in the universe is the same as in a meter, or in a decimeter, or in the farthest stellar trajectory" (*BR,* p. 67). Borges sees in Cantor's proposition a way to escape the bounds of finitude, for if any part, however minute, of an infinite set can also contain infinity, infinities can be opened within any element, no matter how small, simply by suggesting that it is part of an infinite series. "If the universe consists of an infinite number of terms," Borges writes, "then it is absolutely capable of an infinite number of combinations, and the need for a recurrence is invalidated" (*BR,* p. 67).

In its immediate context, this result enables Borges to refute Nietzsche's doctrine of Eternal Return. In the larger context of Borges's stories, Cantor's results seem to promise that a fiction composed of a limited number of words can, like the subsets of an infinite set, nevertheless contain infinity. In "The Aleph," Borges appropriates Cantor's nomenclature and methodology to explore the implications for literature that obtain when infinity is encapsulated within a finite boundary.

The bounded but infinite topography of the Aleph, a sphere a little over an inch in diameter that contains "all space . . . actual and undiminished" (*A,* p. 26), contrasts with the bounded and finite topography of the Garay Street house. The narrator's fixation on Beatriz Viterbo suggests that the arrangement of the locale is a parodic mirroring of Dante's *Divine Comedy,* containing, like that larger topographical work, a three-tiered structure—defined, in "The Aleph," by the cellar, the drawing room, and the modern salon-bar next door.[17] Borges subjects the work of his Italian predecessor to a number of playful inversions: the narrator experiences his epiphany in the lowest, rather than the highest, realm; the Aleph resides in the cellar, while the salon-bar is a horror of modern decor. But these inversions only help to highlight the assumptions that the narrator and Dante share. In using topographical symbolism, both Dante and the narrator assume that space can be assigned symbolic significance, and that this significance can be compressed into, and expressed through, an object in the center of that space. In one sense,the object at the center of that space is Beatrice/Beatriz; in another sense, it is the poem/fiction itself. The parody thus acts to link the narrator's literary strategies with the large body of traditional literature that, like the *Divine Comedy,* aspires to express the ceaseless flux of the infinite universe through a symbolic structure that is itself bounded and finite.

The existence of the Aleph, however, forces the narrator (and us) to question these assumptions. Daneri reverses Borges's procedures. Daneri has no need to resort to symbols to express the essence of the whole, because he is the possessor of an object that literally contains it. He attempts to create a literary counterpart to the Aleph in his poem entitled, appropriately, "The Earth," in which he intends to "set to verse the entire face of the planet" (*A,* p. 19). With the Aleph rather than Virgil as his guide, he believes that literature operates like transfinite mathematics, which implies that the part is as large as, and therefore can contain, the infinite whole. Borges warns that the use of the word "Aleph" for the "strange sphere in my story may not be acciden-

[17]The Dante parody is discussed at length in Alberto J. Carlos's "Dante y 'El Aleph' de Borges," *Duquesne Hispanic Review,* 5 (1966), 35–50; see also Stelio Cro, "Borges e Dante," *Lettere Italiane,* 20 (1968), esp. 407–410.

tal";[18] "for Cantor's *Mengenlehre,* it is the symbol of transfinite numbers, of which any part is as great as the whole."[19] As Borges recognizes, Cantor's work can, if applied to literature, have serious implications for literary methodology. Daneri's procedure illustrates them.

For Daneri, poetic creation consists of putting a subset of reality, the series of signs that comprise language, into one-to-one correspondence with the larger, infinite set of the world. He bases his method on the hint provided by the Aleph. The "formal perfection and scientific rigor" (*A,* p. 21) that Daneri claims for his poem is perfectly correct, if we assume that the world is an infinite set and language one of its subsets, for as we saw, Cantor proved that a subset of an infinite set is as large as the set itself. It is therefore theoretically possible to put the verbal signs into one-to-one correspondence with the infinite sets of the elements that comprise the world. This possibility drastically alters the role that artistic choice plays in literary composition, for when literature is a series of signs to be put into one-to-one correspondence with the world, choice is reduced to deciding which sign to match with which object. Daneri's fickleness in searching for the right sign only emphasizes how unimportant choice has become.

The narrator's fixation on Beatriz, by contrast, renders choice absolute. Rather than being reduced to a minimum as it is in Daneri's poem, choice for the narrator is the mysterious mechanism whereby the infinite multitude of other terms in the world can be made less significant than the privileged one upon which his choice operates. We learn in the course of the story that Beatriz was frail and stooped; that she wrote incestuous, pornographic letters to her cousin; that she was forgetful, distracted, and contemptuous, with a "streak of cruelty" that perhaps "called for a pathological explanation." But these imperfections in no way undermine the power of the artistic choice that identifies her as an ideal love-object. On the contrary, by testifying to the inexplicability of this choice, Beatriz's imperfections emphasize its absoluteness.

[18]The other significance Borges claims for the Aleph is that "for the Kabbala, that letter stands for the *En Soph,* the pure and boundless godhead" (p. 29). This aspect is discussed in Salomon Levy's instructive article, "El *Aleph,* símbolo cabalístico, y sus implicaciones en la obra de Jorge Luis Borges," *Hispanic Review,* 44 (1976), 143–161.

[19]Quoted from p. 29, *The Aleph and Other Stories.* The full title for Cantor's article is "Beiträge zur Begründung der transfiniten Mengenlehre." Part 1 appeared in *Mathematische Annalen,* 46 (1895), 481–512, and part 2 in the same journal, 49 (1897), 207–243. An English translation can be found in *Contributions to the Founding of the Theory of Transfinite Numbers,* trans. P. E. B. Jourdain (Chicago: Open Court, 1915).

This emphasis on choice is significant, for central to the debate that arose over Cantor's transfinite numbers was his method of arbitrarily selecting one element from each of several sets and using them to form a new set. This procedure became known as the "Axiom of Choice," and has since become one of the most controversial axioms of modern mathematics. The importance of the Axiom of Choice to Cantor set theory may be one reason why Borges makes the role of choice in literature an issue in the conflict between the narrator and Daneri.

It is an interesting footnote to Borges's story that in the early 1900s the Axiom of Choice was used to develop the Banach-Tarski paradox, which states that given any two solid spheres (one, for example, the size of a golf ball and the other the size of the earth), each can be divided, as Morris Kline explains, into a "finite number of non-overlapping little solid pieces, so that each part of one is congruent to one and only one part of the other."[20] This implies, as Kline points out, that "one can divide the entire earth up into little pieces and merely by rearranging them make up a sphere the size of the ball."[21] I hasten to remind the reader that this is not Borges's fancy, but a logical consequence of the Axiom of Choice as it is used in modern set theory.

Given these results, it is not surprising that the Axiom of Choice is as central to Borges's story as it is to Cantor's theory. We have seen that for Daneri, choice is simply a matter of matching words to objects in one-to-one correspondence; in this he follows Cantor's own methodology. For the narrator, by contrast, choice is the mechanism whereby the world as a series is first negated by denying its plenitude, then reconstituted (in a form congenial to his temperament) from the single element favored by his choice. Far from using choice to validate series, then, the narrator uses it to deny series. The narrator's antipathy to series is as clear as Daneri's infatuation with them. Whereas Daneri delights in timetables, bulletins, and other paraphernalia that emphasize the seriality of time, the narrator is "pained" by the realization of a "wide and ceaseless universe" in which every small change is "the first of an endless series" (*A*, p. 15).

These different views invite the question, which is closer to Borges's

[20]Kline, pp. 269–270. For the example of the golf ball we could, of course, substitute Borges's Aleph, a sphere "an inch in diameter."

[21]Ibid., p. 270.

own preferences? Borges's strategy follows neither Daneri's one-to-one correspondence method nor the narrator's symbolizing, though it draws on both. Daneri admits the plenitude of the world but does not see that choice is inevitable; the narrator admits choice, but denies the inevitability of the "endless series." Borges can have the best of both methods because he is willing to give up the one assumption that Daneri and the narrator share: that the world exists, and can be represented in literature. Illusion and symbol are therefore admissible, since he has surrendered Daneri's claim to rigorous correspondence; and infinitude is also possible, since he is not bound to the narrator's desire for a single, determinate locus for reality. The Aleph for Borges is not reality but a symbol of the kind of paradox that reveals the impossibility of ever representing reality.

In this distinction lies a crucial difference between Borges and Cantor. Even though Cantor was never able to demonstrate that the paradoxes inherent in his set theory could be resolved, he deeply believed that they *were* resolvable. As the attacks on transfinite number theory grew, Cantor retreated to Platonism for his defense. To Cantor, the alephs were valid mathematical entities because they were ideal objects in the Platonic realm of ideas. Cantor's scientific biographer Joseph Warren Dauben describes how, in Cantor's famous "Mengenlehre" paper, Cantor argued that the paradoxes of his set theory were resolvable because they corresponded to a Platonic reality that was itself consistent.[22] In the ultimate, Platonic sense, Cantor believed that his theory was true because the Alephs were real.

Borges, of course, believes no such thing. In his *History of Eternity,*

[22]Joseph Warren Dauben, *Georg Cantor: His Mathematics and Philosophy of the Infinite* (Cambridge: Harvard University Press, 1979). Dauben calls the sentence with which Cantor begins the "Mengenlehre" treatise a "classic," and adds that it set "the tone for all that was to follow": "*Definition:* By a 'set' we mean any collection *M* into a whole of definite, distinct objects *m* (which are called the 'elements' of *M*) of our perception [*Anschauuing*] or of our thought" (Dauben, p. 170). By defining a set as a collection of elements of our thought, Cantor was able to avoid the distinction that his critics sought to make between the merely abstract existence that transfinite numbers had and the referential existence that various elements within the set may possess. "This characteristic [of the elements in the set as objects of thought] was tremendously important to Cantor in an ontological way," Dauben comments. "If all the elements of his set theory existed on the same level, with the same reality of thoughts and images of the mind, then there was no dependence upon real objects of any sort. . . . The reality of sets as abstract objects in the mind then carried over directly to the transfinite numbers, and conferred upon them a similar sort of reality" (Dauben, p. 171).

Borges impishly suggests that the Platonic realm of eternal archetypes is a huge museum of dusty pieces that serves mostly to frustrate the cabinetmakers of the world as they pursue the unreachable Platonic table.[23] Because Borges does not accept the Platonic reality as "real,"[24] he is also not obliged to accept Cantor's belief that the paradoxes of self-referential systems can be resolved. The Alephs interest Borges not because they are real, but because they allow him to suggest that nothing is "real." This difference in strategies suggests parallel differences in motivation. Whereas Cantor wanted to extend logical analysis, not destroy or compromise it, Borges wants to use logical analysis to show how profoundly illogical its results can be.

The Aleph thus finally means something very different for Borges from what it means for Georg Cantor.[25] Cantor chose the aleph to represent his transfinite numbers because it is the first letter of the Hebrew alphabet, and he hoped that his alephs would be a new beginning for mathematics.[26] In Borges's story, the Aleph also represents fresh beginnings. But Borges is aware that it is a beginning that threatens to burst out of, rather than extend, logical analysis. When the narrator's egoistic eye ("I") attempts to establish bounds around the Aleph by looking at it from "every point and angle," he sees in it "the earth and in the earth the Aleph and in the Aleph the earth . . ." (*A*, p. 28), in a progression that circles back on itself to form a Strange Loop that includes the narrator within its circumference. According to the

[23]Jorge Luis Borges, *Historia de la Eternidad,* vol. 1 of *Obras Completas* (Buenos Aires: Emecé Editores, 1953), p. 21.

[24]In an interview with Ronald Christ, Borges commented that "I think I'm Aristotelian, but I wish it were the other way [i.e., Platonist]. I think it's the English strain that makes me think of particular things and persons being real rather than general ideas being real." *The Paris Review,* 40 (1967), 162.

[25]Borges himself may have underestimated this difference; if so, some of his misunderstanding of Cantor's position can undoubtedly be traced to Kasner and Newman. The book is written in a popularized, colloquial style that might well make other mathematicians blanch. For example, this is their treatment of the very complex issues raised by what "existence" means in mathematics: "In modern times, the various schools of mathematical philosophy, the Logistic School, Formalists, and Intuitionists, have all disputed the somewhat less than glassy essence of mathematical being. All these disputes are beyond our ken, our scope, or our intention. A stranger company even than the tortoise, Achilles, and the arrow, have defended the existence of infinite classes. . . . A proposition which is not self-contradictory is, according to the Logistic School, a true existence statement. From this standpoint the greater part of Cantor's mathematics of the infinite is unassailable" (p. 62).

[26]Kline, p. 270.

narrator, a wrecking firm owned by Zunio and Zungri finally tears down the Garay Street house. The impact of the Aleph lingers beyond the termination suggested by these "Zs," however, and the narrator in a last desperate attempt to break out of the Aleph's Strange Loop decides that the "true" Aleph is buried inside a stone pillar at the mosque of Amr where it cannot be seen, only its "busy hum" perceived. Along with the perfect futility of this final choice goes the absurdity of the narrator's desire to choose a single Aleph as the "real" one. As Borges knows, in Cantor's theory there is not one but an entire succession of Alephs, each no more or less "real" than the last. Borges's fiction implies that the Aleph, like the infinite series from which it derives its name, is a beginning without a terminus, a self-referential object capable of resisting all attempts to define or encapsulate it.

The Aleph thus provides Borges with a metaphor through which he can subvert Cantor's hope that infinity would finally be tamed and brought within the bounds of rational analysis. We have seen that the same shift in perspective that allows Cantor to treat an infinite set as a "transfinite number," seeing it as a coexisting single entity, also brought in its train paradoxes of self-referentiality that finally threw all of analysis into question. Although Borges may not be fully aware of the irony of Cantor's position within the history of mathematics, an intuitive understanding of the potential conflict permeates "The Aleph," where the scientific model is playfully represented through a fictional creation. For Borges, the science is as much a fiction as Daneri's poem, or indeed as the story itself.

I have been suggesting that Borges's response to the field concept is essentially subversive, aimed at revealing the "crevasses of unreason" that make manifest the fictionality of the concept and, by implication, of the holistic reality it tries to express. But Borges, speaking from within that world view, finds his own utterance drawn into question by his subversive strategies. To speak is to engage in sequential analysis and expression, and hence to contradict the simultaneity that is essential for Borges's paradoxes to emerge. Borges fully recognizes this limitation, as the companion essays of "New Refutation of Time" demonstrate. These essays explore the strategy of subversion not merely as an arbitrary or idiosyncratic response, but as a necessary consequence of the field concept that finally subverts the subverter.

In "New Refutation of Time" Borges turns from the possibilities that infinite series have for space to the implications they have for time. He wishes to show that the conclusions of Berkeley and Hume may be extended to deny time. According to Borges, Berkeley denies that there is an object existing independently of our perception of it; Hume denies that there is a subject perceiving the object. With "man" merely a collection of sensations, Borges asks whether a single repeated perception is not enough to deny time also. If the number of human experiences is not infinite, then it follows that perceptions will be repeated, either in one man's life or in the experience of two different men. The repetition has the effect of destroying time as a linear sequence and hence, if time is thought of as a series, of refuting time.

In a now-familiar ploy, Borges makes an almost imperceptible shift from "continuity" to "series" when he talks of time's flow. Borges first refers to the "continuity that is time"; but later he shifts to the "series" of time, as in the following passage: "The metaphysics of idealism declare that it is risky and futile to add a material substance (the object) and a spiritual substance (the subject) to those perceptions. I maintain that it is not less illogical to think that they are terms of a series whose beginning is as inconceivable as its end" (*OI,* p. 176). Though the thrust of the passage is to say that the temporal series is "illogical," its covert effect is to postulate that time is indeed a series rather than a continuity, a series "whose beginning is as inconceivable as its end." Having posited time as a series, Borges then attempts to show that the series is invalid, because perceptions—that is, terms within the series—are repeated. "Is *a single repeated term* enough to disrupt and confound the series of time?" Borges asks (*OI,* p. 178).

In the second essay, the language returns insistently to the terminology of series. Borges instances Chuang Tzu who dreamed he was a butterfly and, awakening, did not know if he was a man who dreamt he was a butterfly or a butterfly dreaming he was a man. Borges maintains that at the moment of awakening, "only the colors of the dream and the certainty of being a butterfly existed. It existed as a momentary term of the 'bundle or collection of different perceptions' which was, some four centuries before Christ, the mind of Chuang Tzu; they existed as term *n* of an infinite temporal series, between n + 1 and N − 1" (*OI,* p. 184). Borges argues that if, "by a not impossible chance," a disciple of Chuang Tzu had an identical dream, the series would be confounded, its progression disrupted by the unexpected repetition of terms. "Is not one *single repeated term* enough to disrupt and confound the history of the world, to tell us that there is no such history?" (*OI*, p. 185), he asks in words nearly identical to those of the first essay.

In order for Borges's argument to succeed, it is *necessary* for him to postulate time as a series. Because time is presented as an unvarying, absolute series, Borges can undermine the relations between terms by suggesting, through the common experience of *déjà vu*, that terms may be repeated in unexpected ways. If the relations defining the position of terms within the temporal series are invalidated, the terms become autonomous units whose arrangement is arbitrary. With time an arbitrary succession of unconnected units, man is merely the "bundle or collection of different perceptions" existing at a point along that succession. When the perceptions change, the man changes. "Identity" is an usual sense thus ceases to exist. To share a perception is to become the same person. "Are the enthusiasts who devote a lifetime to a line by Shakespeare not literally Shakespeare?" Borges asks (*OI*, p. 178).

From the foregoing, it appears that Borges is engaged in refuting the Newtonian idea of time. But the very terms he uses to refute it are imbued with the Newtonian world view, because, as the reader will recall, it is in the Newtonian view that time exists as a series of universal moments. Hence the attempt cannot entirely succeed, because the vocabulary of denial is also the language of affirmation. Borges, of course, recognizes the paradox. Having set up time as a series and shown how it can be disrupted, he proceeds to suggest that the series must inevitably be reconstituted.

That the return of the series is inevitable is suggested most deviously by the following passage. "I repeat," Borges says, "there is not a secret ego behind faces that governs actions and receives impressions; we are only the series of those imaginary actions and those errant impressions." "Series" in this sentence means the temporal series of mental perceptions that constitutes "man." But when Borges immediately continues by repeating the word "series," he changes its referent. "The series? If we deny spirit and matter, which are continuities, and if we deny space also, I do not know what right we have to the continuity that is time" (*OI*, p. 175). In this second use of "series," it does not mean temporal progression, which has now become a "continuity," but rather denotes the sequence of terms emerging from the idealist postulates of Berkeley and Hume. The word "series" in this passage acts as a pivot whose changed meanings contain an essential paradox: first "series" means temporal series, then it means the inevitable extension of the idealist argument. The disruption of (temporal) series is thus made to form the final term of the series proceeding from idealist premises. If the temporal series has been refuted, the series of philosophical conjectures has been extended. To deny series in one instance paradoxically creates a new series, with its implicit chronology of successive generations of philosophical argument.

The two sequential essays of "New Refutation" pose the same paradox in structural terms. The two essays largely repeat each other, confounding the expected seriality of argument by making us experience as repetition what we would ordinarily expect to be progression. "I deliberately did not combine the two into one article," Borges writes, "because I knew that the reading of two similar texts could facilitate the understanding of an indocile subject" (*OI,* p. 172). Reading the two essays does indeed "facilitate the understanding," by making us experience the very repetition that forms the basis for Borges's proof that temporal series can be disrupted.[27] But the essays also exist as a series, as essays "A" and "B". Though they are similar enough to evoke the feeling, "I have read this before," they are different enough to suggest a progression in Borges's thought in the two years separating their composition—that is to say, they exist not only as a repetition but as a progressive temporal series.

Colin Butler, in a rigorous analysis of Borges's examples, demonstrates that the same paradox is true of virtually every piece of evidence Borges adduces to prove that time does not exist. Take the case, for example, of Chuang Tzu. Butler writes,

> so hermetic are Chuang Tzu's respective psychic states that he can never be finally certain whether he was a man who dreamed he was a butterfly, or *vice versa.* The fact remains, however, that he was either one or the other; and while his true identity may be questionable, *that* it is so can only be the consequence of an act of recollection, with its implication of change, and therefore of succession; which his example was intended to controvert.[28]

[27]Ned J. Davison also points out the mimetic form of these essays in "Aesthetic Persuasion in 'A New Refutation of Time,'" *Latin American Literary Review,* 14 (1979), 1–4.

[28]Colin Butler, "Borges and Time: With Particular Reference to 'A New Refutation of Time,'" *Orbis Litterarum,* 28 (1973), 157.

Butler argues, to my mind convincingly, that Borges's essay is a "temporary revaluation [of idealist arguments] within the framework of an ontology that remains conventional throughout."[29] The real agenda for Borges's article, Butler hints, is not to extend the idealist argument to refute time, but to create a framework in which Borges can place, with maximum plausibility and effect, the intimation of eternity that he experienced on one particular summer evening in Barracas. So Butler shrewdly guesses that "A New Refutation of Time" is in fact "written backwards, and its initial philosophizing is only indirectly relevant to what comes after."[30]

If we accept Butler's line of reasoning, the paradox that keeps recurring with the destruction and reconstitution of series can be seen as a result of Borges's inability fully to realize the ontological premises for which he himself argues. Borges can arrive at a felt sense of "the inconceivable word *eternity*" (*OI,* p. 180) only by momentarily suppressing the knowledge that the moment he participates in is not a unique event, but one of an almost endless series of moments, most of which are inimical or indifferent to this feeling. Butler points out that essential to the kind of personal experience Borges is describing "is its capacity to exclude . . . its credibility depends ultimately on the success with which Borges suppresses other felt states which militate against it."[31] Borges's strategy in trying to attribute to the moment a unique status is to separate it out from the flow of time; and this can be done only if the continuum of time is first made into a series. Then the sequentiality of the series is denied or stopped so that the single term containing the desired moment can become omnipresent and of infinite duration.

The problem with this strategy is exactly that encountered by the narrator of "The Aleph." Once the person (or moment, or event) is conceded to form a term in a series, its existence implies the inevitability of the other terms, which will sooner or later emerge to push the favored term out of mind. The narrator confesses at the end of "The Aleph" that "I myself am distorting and losing, under the wearing away of the years, the face of Beatriz" (*A,* p. 30). But positing a series is also essential, for if the world is instead perceived as a continuum, there is

[29]Ibid., p. 159.
[30]Ibid., p. 155.
[31]Ibid., p. 160.

no possibility of separating out the chosen element to begin with. The very action that allows choice to operate also assures that the choice will be less than absolute. The circular dialectic, then, is not merely an arbitrary requirement imposed by Borges's skepticism. Rather, it is a profound recognition of Borges's part that "really, what I want to do is impossible" (*A*, p. 27).

We can now appreciate that the same paradox was implicit from the beginning in Borges's use of transfinite set theory. Cantor's proof that the part could be as copious as the whole seemed to offer a mathematically rigorous demonstration that the single term can be made into the whole. In this case, of course, the necessity of succession is overcome because the part has supplanted the whole. But the victory is illusory, for to accept the presence of an infinite series is also to accept that there will be an entire succession of infinities, with aleph-null succeeded by aleph-one, aleph-one by aleph-two, aleph-two by aleph-three . . . The infinite series that allows Borges to replace the whole with the equally copious part at the same time dooms him to the succession that such a series implies. In "New Refutation," this same paradox besets Borges-the-author as it had earlier Borges-the-narrator of "The Aleph." Borges, speaking in his own voice, is forced to recognize that his intimation of eternity is also subject to the "endless series" that so distressed the narrator of "The Aleph." Thus no terminal resting place is possible, not even the apparent terminus of Borges's own skepticism.

The inherent limitations of Borges's attempt to overcome time become explicit in the last paragraph of "New Refutation." "Time is the substance I am made of," Borges comments. "Time is a river that carries me away, but I am the river; it is a tiger that mangles me, but I am the tiger; it is a fire that consumes me, but I am the fire" (*OI,* p. 187). We thus come, by virtue of Borges's dialectic, around again to the realization that to confirm something is to deny it, to disrupt it is to reconstitute it. The pattern is circular.

As we have seen, the general progression of Borges's argument in "New Refutation" also forms a circle: first time as a series is created; then it is disrupted; finally it is reconstituted, bringing us back to the starting point of the cycle. This recalls the circular patterns imposed on linear series in such stories as "The Circular Ruins" and "The Approach to al-Mu'tasim." But the result now is not merely the loss of the rationality of the Newtonian world view, but the undermining of Borges's own skepticism and idealism. His strategy of claiming that there are no final answers is not final, either.

I began this chapter by comparing Borges with Nabokov and suggesting that their artistic responses to the field concept were fundamentally different. I should like to close by suggesting ways in which they are the same. We have seen that Nabokov took from the new physics the assurance that time can be reversed, placed it in the context of mirror symmetry, and then introduced some slight asymmetry that became identified with the recognition that time does, after all, proceed. Borges appropriates from Cantor set theory the idea of an infinite set as a single, pre-existing entity, and then applies the model to time in order to prove that temporal succession must yield to eternity. But like Nabokov, Borges must finally admit some limitations to his artistic project. Thus both Borges and Nabokov seek to appropriate parts of the field concept for their own purposes, and both end by admitting that the appropriation can never be complete.

The reflection suggests that Borges, like Nabokov, remains grounded in the Newtonian world of ubiquitous, omnipresent time that is finally the ultimate series to which both succumb. Borges himself knows that the series he uses to subvert the field concept involves him in paradoxes that he creates and exploits, and to which he also yields. But worse for Borges than being subject to this limitation is to be trapped within a clear-cut world where continuities of logical progression render paradox impossible. In such a world, all the artist can say is, "The world, alas, is real; I, alas, am Borges."

CHAPTER 6

CAUGHT IN THE WEB

Cosmology and the Point of (No) Return in Pynchon's *Gravity's Rainbow*

Whereof one cannot speak, thereof one must be silent.
Ludwig Wittgenstein, *Tractatus*

MORE THAN ANY OTHER writer in this study, Pynchon grasps the full implications of the field concept, including both its promise of a reality that is a harmonious, dynamic whole and the problem it poses of how to represent that reality in the fragmented medium of language. Pynchon's response to this dilemma is to create a text that at once invites and resists our attempts to organize it into a unified field of meaning. *Gravity's Rainbow* is notoriously difficult to read because its complex and recurring allusions constantly tempt the reader to search for, and recognize, the extensive patterns of interlocking images to which the text owes its remarkable coherence and density,[1] while at the same time frustrating this attempt by a variety of techniques that tend to obliterate or contradict the emerging patterns. Any coherent account of this narrative will have to come to grips with this deconstructing

[1]Often this attempt to recognize patterns in the text takes the form of generic classification. For example, Scott Sanders proposes the term "paranoid history" in "Pynchon's Paranoid History," pp. 139–160 in *Mindful Pleasures: Essays on Thomas Pynchon,* ed. George Levine and David Leverenz (Boston: Little, Brown, 1976); Michael Seidel labels it a "narrative satire" in "The Satiric Plots of *Gravity's Rainbow,*" pp. 198–212 in *Pynchon: A Collection of Critical Essays,* ed. Edward Mendelson (Englewood Cliffs, N.J.: Prentice-Hall, 1978); and Edward Mendelson, in his seminal article, "Gravity's Encyclopedia," pp. 161–196 in *Mindful Pleasures,* identifies it as an "encyclopedic narrative." A strong impulse to fit the novel into a discrete category is also apparent in Mark Siegel's "Creative Paranoia: Understanding the System of *Gravity's Rainbow,*" *Critique,* 13 (1977), 39–54.

dynamic.[2] The patterns of *Gravity's Rainbow* tend toward self-obliteration because the focus for the text's anxiety is precisely the cognitive thought that seeks to organize diverse data into coherent patterns. The source of the tension, in other words, lies in the nature of human consciousness itself.

As we saw in Chapter 2, contemporary philosophers of science and linguists have suggested that the act of cognition is not merely a passive observation of a world "out there," but the active creation of a world that is then perceived as separate from the cognitive faculties that brought it into being. The deconstructing dynamic in *Gravity's Rainbow* is put into the service of the same revised perception. As we try to impose on the chaotic surface of the narrative the cognitive patterns that will let us classify and analyze it, we are forced to become aware of the conscious effort that the reconstruction of pattern requires. Then, when we find connections, that is, images or reflections of our own thought processes, one of the patterns we can discern is the danger of self-consciousness. The realization subjects us to a double-bind. The perceived patterns imply that self-conscious cognition is skewing our society and driving us toward destruction; but in order to receive this message, we had to tame the unruliness of the text into cognitive patterns we understand, thus exercising over the text the same kind of control that created the problem in the first place.

If we turn the dilemma around and look at it from the point of view not of the text's patterning but of its obliteration of pattern, we begin to see what role that refractoriness can play. The text's unruliness makes the reader acutely aware that patterns are not merely perceived but constructed, thereby alerting us (or reminding us) that as we read, we are building a cognitive structure. At the same time, the unruliness insures that this cognitive structure cannot be complete or perfect. For all its frustration for the reader, the unruliness offers a way out of the central dangers of authoritarian control and life-denying organization.

I should like to turn now to a discussion of two central "patterns" in *Gravity's Rainbow,* and show how they invite our identification of them

[2]Edward Mendelson's suggestion that the text is an "encyclopedic narrative" has proven among the most fruitful so far because his classification allows him to acknowledge some of this resistance. But the refractoriness of the text goes beyond the qualities (gargantuanism and inclusiveness) that Mendelson identifies as resisting organization. What the text exhibits may more properly be described as a deliberate obliteration of pattern.

as "really there" and at the same time frustrate any unequivocal or unambivalent interpretation of them. In their simplest aspect, the two patterns may be represented as the circle and the line. The circle is consistently associated with natural cycles and processes, and with the prospect that we can Return to some simpler, more innocent identification with the universal "field" of the cosmos. The line, or circle that has become linear by being opened into a parabola, is associated with the artificial structures of control that drive toward some final terminus.[3] The contrast is embodied in two forms similar in shape but antagonistic in meaning: the rainbow and the ballistic arc of the Rocket. Whereas the circular rainbow descends onto the "green wet valleyed Earth" in a harmonious and fertile union, the parabolic Rocket's arc, Katje senses, is a "clear allusion to certain secret lusts that drive the planet and herself, and Those who use her."[4] To create the "visible," upward part of the parabola whose arms stretch downward into infinity, the Rocket must thrust against Gravity, and this ascension can be achieved only through the naked application of power. It "alludes" to the ethos of patriarchy, the search for power that identifies masculinity with technology. Yet there necessarily follows Brennschluss, the point at which the Rocket enters its "feminine" aspect, when the assertion of power is helpless in the inevitable submission to a patriarch still more ancient, "Old Gravity." When the Rocket enters its descent, it "has submitted. All the rest will happen according to the laws of ballistics. Something else has taken over. Something beyond what was designed in" (p. 223). The Rocket's arc, unlike the rainbow, thus has a fatal asymmetry, the assertion of power at the beginning inevitably leading to the destruction at the end of the descent, the Rocket prefiguring the planet's "plunging, burning, toward a terminal orgasm" (p. 223).[5]

[3]Alan J. Friedman and Manfred Puetz, in "Science as Metaphor: Thomas Pynchon and *Gravity's Rainbow*," *Contemporary Literature*, 15 (1974), 345–359, present an excellent discussion of the cycle of life and death as a continuing process. An implication of the cycle, they conclude, is that order and chaos are mutually entailing opposites.

[4]Thomas Pynchon, *Gravity's Rainbow* (New York: Viking Press, 1973), p. 233. I indicate my ellipses in brackets; when they appear unbracketed, they are Pynchon's.

[5]Pynchon's scenario may well be indebted to the principles Freud tentatively sets forth in "Beyond the Pleasure Principle." Musing on the "compulsion to repeat" found in both neurotic and normal behavior (p. 22), Freud hypothesizes that the urge toward repetition is the desire to return to an earlier state of being. The earliest state of being is, of course, the inanimate. Freud speculates that this drive is controlled by a set of "conservative" instincts, and that these are perhaps even stronger than the pleasure-seeking ones. He is thus "compelled to say that '*the aim of all life is death*' and, looking backwards, that

The opposition between two geometries, one pointing toward Return, the other skewed by the self-conscious desire for control toward a terminus from which there is no Return, is repeated in the contrast between the synthetic molecules of the chemical cartels and the organic, fertilizing molecules of nature. The narrator recounts how in Kekulé von Stradonitz's famous dream of 1865, Kekulé sees the Uroborus, the "dreaming Serpent which surrounds the World" (p. 412). The Serpent announces that "the World is a closed thing, cyclical, resonant, eternally-returning" (p. 412); but Kekulé interprets it as a vision revealing the cyclic structure of benzene. Kekulé's discovery of the structure of benzene opens the way for widespread chemical synthesis of organic compounds; thus the dream signifying Eternal Return is perverted into a synthetic parody of itself. The dream has come, the narrator tells us, to a "system whose only aim is to *violate* the Cycle. Taking and not giving back [. . .] removing from the rest of the World those vast quantities of energy to keep its own tiny desperate fraction showing a profit [. . .]. The System may or may not understand that it's only buying time [. . . that] sooner or later [it] must crash to its death, when its addiction to energy has become more than the rest of the World can supply, dragging with it innocent souls all along the chain of life" (p. 412).

In contrast to this perversion of Return is the process of decay that fertilizes Pirate Prentice's rooftop garden. Here the cycle of Return, mirrored in the cyclic form of the molecules, signifies not synthesis and death but fertility and life. The "politics of bacteria, the soil's stringing of rings and chains in nets only God can tell the meshes of," produce bananas a foot and a half in length, "yes amazing but true" (pp. 5–6). The affirmation of life continues as the enchanting odor of the banana breakfast takes over "not so much through any brute pungency or volume as by the high intricacy to the weaving to its molecules [. . .] it is not often Death is told so clearly to fuck off" (p. 10). The "living genetic chains" that prove "labyrinthine enough to preserve some human face down ten or twenty generations" also create the "same assertion-through-structure [that] allows this war morning's banana

'inanimate things existed before living ones.'" The Standard Edition of the Complete Psychological Works of Sigmund Freud, trans. James Strachey, vol. 18 (London: Hogarth Press and the Institute of Psycho-Analysis, 1955), p. 38. It is possible, of course, that the Freudian influence is mediated, as Lawrence C. Wolfley suggests, through Norman C. Brown.

fragrance to meander, repossess, prevail" in spite of the terror of falling bombs. In contrast to the artificial structures of organization and control that deny the cycle of Return, then, are the natural structures of decaying organic matter that embody and affirm it. One is evil and insane, driving toward death; the other is natural and good, a source of life and hope.

The comparison reveals that it is not simply structure in itself that is the source of the narrator's concern, because the natural world is also ordered. Rather, the concern is specifically with those structures that allow Them to consolidate control by extending the images of human consciousness to all creation. The natural structures, by contrast, reveal creation as a unified field in which we as well as all other creatures participate, but which does not specifically valorize consciousness. Pynchon's sense of the field concept, derived largely from thermodynamics and cosmology, emphasizes that it can never be entirely captured or contained within cognitive structures.

Pynchon's suspicion of cognition can be seen in his treatment of the "Other Side," the existence we pass into after we die. When the narrator is in the grip of paranoia, he imagines that the Other Side, like This Side, has been corrupted by cognition into organizational structures of control. When he can temporarily shake off his paranoia, however, he imagines the Other Side as an initiation into a field view of reality. Crossing to the Other Side can convert even the most confirmed bureaucrat into a prophet of the field concept. Lyle Bland, for example, is transformed from plutocrat to mystic when he stumbles on the Masonic ritual whose magic allows him to have out-of-body experiences. A similar transformation occurs with Walter Rathenau. Rathenau, though he was "prophet and architect of the cartelized state" on This Side (p. 164), once on the Other Side begins to "see the whole shape at once," and as a result looks at the cartel in a radically different way. "Let me be honest with you," he says to the German technocrats who are his successors. "I'm finding it harder to put myself in your shoes" (p. 165). So changed is Rathenau from those who are "constrained, over there, to follow it [the pattern of history] in time, one step after another," that he does not hesitate to tell the assembled bureaucrats that "if you want the truth [. . .] you must ask two questions: First, what is the real nature of synthesis? And then: what is the real nature of control?" "You think you know," he warns them, "you cling to your beliefs. But sooner or later you will have to let them go" (p. 167).

As Rathenau suggests, the issues of control and synthesis are central to understanding our perversions of the field view. In Kekulé's dream, the bureaucrats on the Other Side try to bring dreams within Their control. But in Rathenau's seance, control itself is revealed as a dream. The bureaucracy aims to reinforce pre-existing patterns of consciousness by reassuring the self-conscious mind that even its moments of unconsciousness are controlled by "switching-paths"; but the spirits from the Other Side insist that even moments of supposed self-awareness are an illusion.

The illusion of control leads inevitably to the attempt at synthesis. To maintain control we synthesize molecules, cause-and-effect, and language, all of which symbolize and complete our alienation from the natural world. Roland Feldspath tries to articulate what a field view means when he tells his listeners that from Beyond he can perceive the "illusion of control. That A could do B. But that was false. Completely. No one can *do*. Things only happen, A and B are unreal, are names for parts that ought to be inseparable" (p. 30). "All talk of cause and effect is secular history," Rathenau says, "and secular history is a diversionary tactic. Useful to you, gentlemen, but no longer so to us here" (p. 167).

As we shall see, the solution to the problem is by no means simple or unambiguous, since it implies abandoning our very identity as *homo sapiens,* the creature who knows. Yet this is nevertheless the transformation that the narrator seems to suggest we should undertake. Cognition is to be replaced by an appreciation for the synchronicity of Nature in which humans do not stand outside as originators of pattern, but take their place within already existing harmonies that include and subsume them. What I have been calling the narrator's obliteration of pattern may thus be an attempt to create another kind of pattern, a pattern that is holistic rather than sequential, synchronous rather than causal, natural rather than specifically human.[6]

Crucial to this enterprise is Pynchon's narrative technique, as we can

[6]An exposition of synchronicity as a world model is articulated by C. G. Jung in *Synchronicity: An Acausal Connecting Principle,* trans. R. F. C. Hull, Bollingen Series vol. 20 (Princeton: Princeton University Press, 1973). In developing his theory, Jung was greatly influenced by quantum mechanics and the Uncertainty Principle through correspondence with Wolfgang Pauli, who was one of his patients.

see by comparing it to more traditional novels. The traditional novel narrates action that we take to be meaningful, but it also contains descriptive passages whose purpose is to set the scene. This "loose bagginess" of the realistic novel has the effect of valorizing our usual modes of cognition, which depend upon subordinating masses of detail into background that can be safely ignored while we concentrate on the small area brought into focus by our conscious attention. Psychologists have shown that this subordination of perceptual data into background is an essential element of cognition; it is what allows us to "tame" the incoming signals so that we are not constantly overwhelmed by a mass of detail.[7] The texture that we identify as "novelistic" recapitulates this process by encoding its signs with distinctions between significant events and "irrelevant" details. The traditional novel is thus "realistic" precisely in the sense that it mirrors the process that allows us to bring reality into focus.[8]

What the familiarity of these conventions may keep us from seeing is that the apparently "irrelevant" details perform an indispensable function. To use a painterly analogy, the "irrelevant" details define the background against which a figure can be discerned. Though the figure is usually the focus of our attention, the contrast with the background is what makes the figure visible. Meaning is made possible—indeed, defined—by its distinction from non-meaning.

Pynchon explodes the traditional distinction between foreground and background by taking a radically egalitarian attitude toward his material. In effect, he refuses to make the distinction between the meaningful event and the "irrelevant" detail. In *Gravity's Rainbow* everything is placed at an equal distance from the reader, so that background and foreground collapse into the same perceptual plane. *Gravity's Rainbow* has hardly any plot in the conventional sense of the word—or rather, more accurately, everything is plot. Flashbacks, bits of dialogue from other times and places, sudden shifts of scene and personae, startling

[7]For a nontechnical account of this process, see M. Toda, "Time and the Structure of Human Cognition," in *The Study of Time II*, ed. J. T. Fraser and N. Lawrence (New York, Heidelberg, Berlin: Springer-Verlag, 1975), pp. 317ff.

[8]The constructive role of cognition is stressed by Ulric Neisser in *Cognitive Psychology* (Englewood Cliffs, N.J.: Prentice-Hall, 1967), p. 3: "the proximal stimuli bear little resemblance . . . to the object of experience that the perceiver will construct." Cognition is the sum of those processes that create from the incoming sensory data a recognizable picture of the world.

transformations of apparently realistic scenes into bizarre fantasy sequences, make of the narrative a mosaic that defies the attempt to see meaning by making a series of distinctions, because all are treated by the narrator as equally valid, equally entitled to our attention. With everything at an equal distance from the reader, meaning can no longer be achieved by discerning difference. Rather, it emerges as a result of seeing that everything is at the same distance. Meaning thus becomes a function not of difference but of similarity, arising not from distinguishing parts but from seeing the interconnection of the whole.

Pynchon's equalizing technique, a kind of "principle of equal distance," establishes a link between our experience of the text and his central themes. We tend to experience meaning in this text as a paranoid, or someone dropping acid, or a religious visionary who believes in Providential design might experience it; in these views, there are no irrelevancies. These very different ways of organizing experience are isomorphic in the sense that they all presuppose the pervasiveness of pattern—that is to say, they all suppose a field view of reality.

The changed way in which meaning arrives helps to explain why readers react to this book in singular ways. They tend to be divided between those who find the novel a chaotic mass of unconnected detail, and those who see its patterning as pervasive. The difference in perspective arises because in *Gravity's Rainbow* meaning arrives as a *gestalt,* precipitating into awareness; either one sees the whole design, or one doesn't see it at all. For those who do, the technique forges a bridge between the emerging sense of a field view and the experience of reading. The very fact that we can see the connections means that we are participating in the mode of vision being described. *Gravity's Rainbow* is thus both a narrative and an initiation.

The problems of articulation that the field concept presents are not, however, so easily escaped, because of the sequential way in which human perception proceeds. To illustrate, consider the "reversible" drawings that psychologists call equivocal figures. These are black-and-white pictures that can appear either as a white figure on a black ground or as a black figure on a white ground. Once we learn to see both figures, we can bring first one, then the other into focus by shifting our attention from one figure to the other. It is apparent that one's mindset, rather than the picture itself, determines which figure is primary. Because neither could exist without the other, to designate either as ground or figure is arbitrary; they mutually define each other. The relevant point is that human cognition is such that only one figure can be brought into focus at a time, even though we know that they are interconnected.[9] As a consequence, we can never see the whole picture at once, because half of it will always be obscured from our vision at any given time.

This limitation is sensed by the lone visionaries of *Gravity's Rainbow,* who try to tell the world that what we perceive as opposites are really parts of an interconnected whole. William Slothrop, a distant ancestor of Tyrone, is one of these. As he drives his beloved pigs to market, he begins to see that the joy of the journey is defined by the slaughter of pigs at the journey's end; the "squealing bloody horror at the end of the pike was in exact balance to all their happy sounds" (p. 555). If communion and slaughter are mutually defining opposites, then moral judgments that label one as "evil," the other as "good," miss the point of their mutual entailment. In his tract *On Preterition,* William Slothrop extends the argument to refute the Puritan doctrine of the elect. There could be no elect without the preterite, William reasons, since salvation makes sense only if there is also damnation. The preterite are thus as necessary to the scheme of things as their betters and hence, paradoxically, as worthy to be saved.

To realize that the distinction between the elect and the preterite implies a fragmentation of an original unity is to attain some hint of the natural wholeness that precedes and pre-empts such dichotomies. Those who, like William Slothrop, can realize this fallen condition have the best chance of wending their way back to a holistic mode of vision. Pointsman scorns the conjunction of opposites as "just this sort of yang-yin rubbish" (p. 88), and he stands for those positivists who are wedded to the atomistic perspective. William Slothrop, by contrast, represents the "fork in the road America never took, the singular point she jumped the wrong way from" (p. 556).

To Pynchon, the theme of Return means more, then, than just returning in time or allowing a cycle to continue. It also means learning to see the implications of the field concept, learning to appreciate the

[9]That we focus on only one aspect of an equivocal figure at a time is emphasized by E. H. Gombrich in the introduction to his classic work, *Art and Illusion: A Study in the Psychology of Pictorial Representation,* Bollingen series vol. 25 (Kingsport, Tenn.: Pantheon Books, 1960), pp. 5–7. An example of an equivocal figure can be found in Neissen, p. 90.

many levels of the realization that "everything is connected." Even so megalomaniac a character as Gerhardt von Göll senses the profound truth that Pynchon believes the *gestalt* view contains. Greta Erdmann recalls that when Gerhardt filmed *Alpdrucken,* he arranged the lighting so that it "came from above and below at the same time, so that everyone had two shadows: Cain's and Abel's" (p. 394). Lest we dismiss the conceit as just so much "precious Göllerei," the narrator later assures us that the "Double Light was always there, outside all film, and that shucking and jiving moviemaker was the only one around at the time who happened to notice it and use it, although in deep ignorance, then and now, of what he was showing the nation of starers" (p. 429). Later, in the guise of Der Springer, Gerhardt tells Slothrop, who is feeling sorry for the preterite crowd longingly eyeing Närrisch's dead turkey, "Be compassionate. But don't make up fantasies about them. Despite me, exalt them, but remember, we define each other. Elite and preterite, we move through a cosmic design of darkness and light, and in all humility, I am one of the very few who can comprehend it *in toto*" (p. 495).

If we "in all humility" try to comprehend the "cosmic design of darkness and light" that *Gravity's Rainbow* is weaving, we can begin to see how the field concept encourages thematic connections between what at first appear to be complete contraries, as different as black and white. Death, for example, is sometimes represented as black, like night, like shit, like the Hereroes. At other times it is Dominus Blicero, that which whitens, like Weissmann, like the Imipolex shroud that Gottfried wears as he hurtles to his death. Drawing on Lawrence C. Wolfley's work on repression in *Gravity's Rainbow,*[10] we can appreciate the complex of meanings that unites these black and white aspects of death into a *gestalt*.

In its black aspect the *gestalt* points to a repressed complex of emotions, beginning with the white man's unconscious association of death and shit, a "stiff and rotting corpse [. . .] inside the white man's warm and private own *asshole*" (p. 688). When the white man associates the black and brown races with shit/death, the repressed complex manifests itself as racism or, in times of more acute stress, as the systematic

[10]Lawrence C. Wolfley, "Repression's Rainbow: The Presence of Norman O. Brown in Pynchon's Big Novel," *PMLA,* 92 (1977), 873–889.

genocide of people of color. In its white aspect death is associated with the white bureaucrats and their attempt to routinize death by routinizing life. Their denial of the black aspect of existence can be achieved only at the expense of making life not white, but gray. The resulting desire to inject some color into a colorless life leads to such aberrations as Weissman's games, on one level, and Pökler's compliance with the technology of death on another. Meanwhile, the technocrats continue to exploit us, and every other life-form, merely to forestall for a little longer the inevitable. Pynchon is thus able to link his recurring concerns—the rise of the multinationals, racism, the arms race, an exploitive technology—with the field concept, which at once clarifies the thematic connections and forces us into a new mode that is one version of the path that, somewhere back in the sixteenth or seventeenth century, we "jumped the wrong way from."

At this point it may be well to clarify how the field view differs from the kind of connectedness They seek. In one of Pirate Prentice's fantasies, he envisions the preterite hell reserved for double agents, where Father Rapier, a "devil's advocate," preaches "like his colleague Teilhard de Chardin" against Return (p. 539). In *The Phenomenon of Man,* Teilhard de Chardin argued that the evolutionary process would continue, with humans evolving into higher and higher levels of cognition, becoming more and more highly conscious.[11] In Father Rapier's speech, this theory takes the form of an assertion that "once the technical means of control have reached a certain size, a certain degree of *being connected,* one to another, the chances for freedom are over for good" (p. 539). Father Rapier sees that this connection leads directly to the possibility that They will not die. The narrator acknowledges that the devil's advocate makes a "potent case" (p. 539); it represents the opposite of true Return, the perversion of wholeness into a kind of connectedness that both derives from and extends consciousness, with the ultimate aim of defeating death.

The difference between Pynchon's sense of the field concept and this perversion of connection is that the field concept makes us aware that the pattern is more inclusive than we can ever see at any one time, that another figure—at present perceived as ground—is also part of the

[11] Pierre Teilhard de Chardin, *The Phenomenon of Man,* trans. Bernard Wall (New York: 1959).

pattern. Although in a sense consciousness brings the pattern into being, the nature of the figure is such that it also reminds us of the limits of consciousness. More important, it is a pattern that includes rather than denies death. By acknowledging both black and white, it admits death as an essential part of the cycle of Return. Though from the point of view of self-conscious cognition death is a termination, in the field view it is the background that, projected into the foreground, completes the pattern.

We have seen how the black and white dualities that are pervasive in *Gravity's Rainbow* can be united into a single equivocal figure by an act of *gestalt* perception that not only reveals the interconnectedness between the apparent opposites, but also initiates us into a field view that is in sharp contrast with the fragmented, atomistic mode of cognitive consciousness. But in order to appreciate this implicit unity, we first have to indulge in some fairly sophisticated cognition, for example the analysis proposed by this chapter. The paradox points to the fact that the enterprise that Pynchon is undertaking cannot succeed, almost by definition. To be able to decode the text is to be cognitively conscious, and to be cognitively conscious is to deny the message implicit in that decoding. We thus return to the double-bind that is at the center of Pynchon's complex view of our relation to the field concept. What then are we to make of the narrative as an attempt to create a verbal analogue to a comprehensive, interactive field? When the attempt to Return is subverted through the very processes the narrative uses to envision it, the enterprise of representing the field through words itself becomes ambivalent.[12] The narrator's divided response to his own narrative project emerges with special force in the metaphors of the frame and the interface.

In *Gravity's Rainbow* we are always in the process of reconstructing, of piecing together the bits and pieces of what we hope will be a complete picture. Yet even to call it a "picture" is to frame it and thus to falsify the attempt to create a holistic vision. The pun on frame is important: by placing the narrative within a frame, we view it as essen-

[12]Linda Westervelt makes a similar point in "'A Place Dependent on Ourselves': The Reader as System-Builder in *Gravity's Rainbow*," *Texas Studies in Literature and Language*, 22 (1980), 67–90, when she points out that *Gravity's Rainbow* "is troubling not only because of the events it portrays, but also because of the activity required of the reader and the comments upon that activity which the text implicitly makes" (69).

tially separate and distinct from the cognitive faculty that brought it into being. Thus we have "framed" it in another sense, that is, defined it in such a way that what we assert of it, even though false, cannot be proven to be false because the falsity is contained in the very assumption that it is an object of discourse.

Another sense of "frame" is the frame of a film, the individual segment that has been artificially imposed on the original scene by the photographic process. The square holes punctuating the narrative resemble the sprocket holes of film, suggesting that the intrinsic discontinuity of film is recapitulated in the fragmented sections of the narrative. When a film is shown, the frames blur together to give the illusion of a continuous process; but the image on the screen is in fact a fragmented reconstruction of what was only an actor's version of reality to begin with. The narrative, like the celluloid strip of film, is an interface that creates through fragmented words (as the film does through color and light) a similitude of continuous reality. If we think of an interface as a barrier inserted between us and the "real thing," we may imagine that if only we could penetrate it, we could get back to the reality whose image we see. But that "reality" is itself a reconstruction. Just as the "reality" of a film derives from actors moving through a script, so the "reality" of the verbal reconstruction derives from cognitive processes that transform immediate perceptions into verbal abstractions. Return by this route is thus impossible, since what we Return to is only another reconstruction, not the original field of life itself. One meaning of the interface, then, is that "reality" exists on neither side, so that penetrating the interface merely takes us from one kind of reconstruction to another.

More generally, an interface is a boundary separating two phases of matter, or metamorphically, two orders of being. In fragmented, pretertite terms, an interface encourages us to see the two phases as qualitatively different, and therefore to perpetuate the illusion that *one* of them, at least, must be real. Thanatz takes this view of the interface between This Side and the Other Side as he waits at the "black and white" gasworks for the Blicero that he suspects may be dead. Thanatz, as death-obsessed as his name suggests, can't resist imagining "what it will take to get Blicero across the interface. What ass-wiggling surrender might bring him back . . ." (p. 668). Thanatz recognizes the interface as a "meeting surface for two worlds," but the narrator adds, with his usual ironic ellipsis, ". . . sure, but *which two?*" (p. 668). To Thanatz the interface leads merely to new opportunities to continue the games of dominance and submission, control and surrender, that self-conscious humanity amuses itself with in a parodic perversion of true Return.

But the interface also hints at a wholeness that cannot be grasped through rational analysis. At the interface, Thanatz confronts these limits in understanding the possibilities for Return. He knows that

> there's no counting on any positivism to save him, that didn't even work back in Berlin, before the War, at Peter Sachsa's sittings . . . it only got in the way, made others impatient with him. A screen of words between himself and the numinous was always just a tactic . . . it never let him feel any freer. These days there's even less point to it. (p. 668)

From a field perspective, interfaces are not barriers, but points of exchange, surfaces through which two orders of being can interpenetrate. This raises the possibility of a holistic field that transcends and includes the interface. As the interface is thus transformed from the boundary that cognitive consciousness perceives it to be to the permeable membrane it can become in the field view, Return returns again as a possibility.

How an interface might become a permeable membrane is suggested by the narrator's treatment of film. When we think of an interface as a barrier, we imagine that on one side of a film are the screen images, the complex play of light and shadow that creates the illusion of life; on the other side, actors performing actions called for by the script. But what the narrator makes us see is that these screen images sometimes have consequences in life beyond the script, as when the jackal men rape Greta Erdmann in *Alpdrucken* and father upon her a real child, Bianca, who will later die in the jackal ship *Anubis*. The interconnections extend to the men watching the film who will that night go home and father children themselves, as Pökler does with Ilse. Connecting white Bianca with dark Ilse is the celluloid film impregnated with silver nitrate salts, chemicals whose peculiar property it is to translate light into black, darkness into light. When Ilse is identified as the shadow image of Bianca, we are once again reminded that, despite our divisions of the field into This Side and the Other Side, art and reality, connections exist that join all sides into a single field of interaction.

In a narrative where black and white are joined by *gestalt* perception into a single figure, the interfaces are always permeable, given the right perspective. Van Göll, hearing of the "accidental" death of the composer Webern, insists that everything fits together and that there are, properly speaking, no accidents. "One *sees how* it fits, ja? learns patterns, adjusts to rhythms, one day you are no longer an actor, but free now, over on the other side of the camera [. . .]" (p. 494). For Der Springer, whose totem is the white plastic knight, this means "waking up one day, and knowing that Queen, Bishop and King are only splendid cripples, and pawns, even those that reach the final row, are condemned to creep in two dimensions, and no Tower will ever rise or descend—no: *flight has been given only to the Springer!*" (p. 494). As the Springer in his imagination leaps off the chessboard, departing from the game in which he is a piece to be moved, an actor reading lines, a character in a book, to "the other side of the camera," we are invited to reflect on the implications of seeing every text or film as a permeable membrane.

The possibility that art can become reality, reality art, is a double-valued potential in *Gravity's Rainbow,* bearing both negative and positive signs. On the one hand it opens the possibility for Return, for if the artist in an act of re-vision can picture the wholeness that we have lost we may be able to recapture it, not only in imagination but in fact. These moments of possibility shimmer throughout *Gravity's Rainbow* like a rainbow of promise. When Geli saves Tchitcherine, the narrator wistfully hopes that "this is magic. Sure—but not necessarily fantasy" (p. 735). The routes back, Pynchon intimates, are real; the Masonic ritual, though debased into something as innocuous as Rotary luncheons, really does work; love as a redemptive force really does have efficacy in the world. At the moment when Geli's spell proves strong enough to counter Tchitcherine's obsession to kill the black brother who is his color-negative, the narrator can hope that the life-force, the forces of Return and redemption, will be stronger than the destructive impulses of a death-haunted humanity. It is not hard to hear in these passages the hopefulness of the revolutionary rhetoric of the 1960s. The death force is "only nearly" as strong as the life force because "a few keep going over to the Titans every day, in their striving subcreation" (p. 720).

This "subcreation" is, however, subsumed within the creation that is *Gravity's Rainbow,* and in that larger context, the narrator's hope that our plastic age can again become titanic is more complex. That the larger act of vision, *Gravity's Rainbow* itself, is a kind of Return is hinted at in the same pervasive pattern that makes of it a *gestalt;* the very wholeness of its design suggests that it is possible to learn to grasp the field view. But Pynchon cannot help also seeing the ironic possibilities of this kind of Return. They come out most clearly, perhaps, in his treatment of von Göll.

After von Göll hears about the Zone-Hereroes, he becomes convinced that his fake propaganda film on the "Schwarzkommando" in fact brought them into being. So he makes contact with the Argentine anarchists who want to restore their land to the primal unity it possessed before the national government and white men made it into a land of broken promises and broken landscape. "It is my mission," von Göll tells the Argentine anarchists, "to sow in the Zone the seeds of reality [. . .] My images, somehow, have been chosen for incarnation. What I can do for the Schwarzkommando I can do for your dream of pampas and sky. . . . I can take down your fences and your labyrinth walls, I can lead you back to the Garden you hardly remember" (p. 388). The allusions to Borges, however, encourage us to put von Göll's promise in a different perspective. As we saw in Chapter 6, Borges does not remove labyrinth walls in his fictions; rather, he crafts his tales with such convoluted turnings that they themselves become verbal labyrinths. We never get back to the "Garden you hardly remember" in Borges. Rather, the Garden is itself revealed as an artifact, a labyrinth of our own making. The same could be said of Pynchon; though the pervasive patterning of *Gravity's Rainbow* compels us to see "everything is connected," what is restored is not primal unity but a postlapsarian artifact that feeds on paranoia and complicity.

In more general terms, the problem is that the vision of the artist is necessarily that of his time and place—fallen. Even granting the power, what sort of creation would this fallen Creator bring into being? The problem reaches grotesque proportions with von Göll; what vision of wholeness could emerge from this mad megalomaniac? The incongruity between von Göll's promise and the reality he represents is immediately apparent in his plans to film *Martin Fierro.* In Part I of the poem, Martin Fierro deserts and turns renegade, abandoning the army to side with the Indians and the open land. But in Part II, the *Return of Martin Fierro,* the gaucho "assimilates back into Christian society" and returns to the city; "a very moral ending, but completely opposite to the first" (p. 387). The circularity implies a failure of artistic vision that makes true Return impossible.

> "What should I do?" von Göll wants to know. "Both parts, or just Part I?"
>
> "Well," begins Squalidozzi.
>
> "I know what *you* want. But I might get better mileage out of two movies, if the first does well at the box office." (p. 387)

As far as der Springer is concerned, the snake in his promised second Garden doesn't need an apple; he can just hold up a balance sheet.

The problem is bigger than the idiosyncrasies of the megalomaniac German filmmaker; it is at the heart of the moral ambiguity that informs Pynchon's vision of the artist as creator. If film is one permeable membrane, the Word is another; and Pynchon, like von Göll, is fallen, preterite, of diseased imagination. What then if his acts of naming, like von Göll's films, have the power not just to reveal the patterns but actually to create them? What rough beast will come breaking through the Text as interface into this world? It is this question that bestows on the act of naming, as on other acts of creation, a terror that cognitive consciousness at once creates and apprehends.

There is, in addition, an ever more radical problem with Pynchon's act of naming that arises not merely from the preterite nature of the artist but from the fallen nature of language itself. When the early Slothrops—Constant and Variable—believed that the Scriptures would be translated directly into God's Hand emerging from the sky, they were testifying to the power of the Word in Western culture. God's Word wrought the first Creation; his Son is the Word become flesh. But in the fallen world of the preterite, the status of the Word as an instrument of creation is more ambiguous. Under narcosis Slothrop comes up with "Blackwords," "new coinages [that] seem to be made unconsciously" (p. 391). Has he, the narrator asks,

> by way of language caught the German mania for name-giving, dividing the Creation finer and finer, analyzing, setting namer more hopelessly apart from named, even to bringing in the mathematics of combination, tacking together established nouns to get new ones, the insanely, endlessly diddling play of the chemist whose molecules are words . . ." (p. 391).

In this view language leads not to connection but to fragmentation or even random combination.

The kind of Creation spelled out by the Word in a routinized society is indicated by the internecine rivalries among the Committees set up to form the New Turkic Alphabet. Representative of these insanities is Radnichny on the Schwa Committee, who "has set out on a megalomaniac project to replace every spoken vowel in Central Asia" with a "neutral *uh*" (p. 353). Pynchon's broad satire points to the imperialism of the entire project, which aims to impose on "the lawless, the mortal streaming" (p. 355) of native speech the letters and words that the bureaucracy decides they should have. Once the flux of the spoken Word has been broken into the discrete and inflexible symbols of written language, it "can be modulated, broken, recoupled, re-defined, co-polymerized one to the other in worldwide chains" (p. 355). Treating words as chemicals, and chemicals as words ("How alphabetic is the nature of molecules," the narrator muses) suggests that their different programs have the same end, control, and the same effect, a numbing, vitiating fragmentation of Meaning into meaningless segments and combinations. The positivist program to force multiform indeterminacies into specific slots, whether through dictionaries, chemical synthesis, or Pavlovian experiments, breaks the original Whole into shards that can then be recombined to reflect the face, not of God, but of fallen man.

In view of the uses and abuses the preterite find for the Word, one can only applaud Slothrop's instinct to "edit, switch names, insert fantasies" (p. 302) in the yarns he spins for Tantivy at the office. Slothrop's primitive fear of "having a soul captured by a likeness of image or by a name" (p. 302) is one expression of the ambiguity inherent in the project of creating words; once written (or published in a book), they inevitably become transfixed, discrete, and immutable, an implicit denial of the ever-changing flux of the field view, even if they purport to embody or reflect that endless streaming.

Thus the moments when the Word becomes potent with the possibility that it will not only describe reality but also bring it into being, moments when "pencil words on your page [are] only Δt from the things they stand for" (p. 510), are charged with threat as well as promise. The possibility that the Text is an interface, on one side the Word, on the other reality, haunts *Gravity's Rainbow*. It is at this threshold, when the characters can feel "the potency of every word," when "words are only an eye-twitch away from the things they stand for" (p. 100), that the ambiguities surrounding the possibility of Return become most apparent and painful. For if the act of naming itself introduces division, what could these moments bring into being but the fragmented reality that cognitive processing implies? And if the fragmentation of that named creation is only another version of Their synthetic, fragmented world, then the whole project of escaping Their control has been co-opted and subverted by the very attempt to speak it.

Pynchon's view of the potency of the Word is subtly different from the Adamic belief that everything possesses its own right name, ordained by God and pronounced by man. Being of the post-structuralist generation, Pynchon grants that names in themselves are arbitrary. But he reasons that once humankind has assigned meaning to sounds, how we choose to deploy those sounds (the *act* of naming) reveals deep patterns of correspondence that the namers sense and to which they respond. So when Säure Bummer grabs the Wagnerian helmet, screws the horns off of it, crowns Slothrop with it and screams, "Raketemensch!" the narrator comments, "Names by themselves may be empty, but the *act of naming* . . ." (p. 366). Enzian, too, when he intuits that the Rocket was fired from Nordhausen, house of the north and therefore of death, thinks that "names by themselves may have no magic, but the *act* of naming, the physical utterance, obeys the pattern" (p. 322). Herero history also "obeys the pattern," so the correspondences between the Herero past and the Rocket are not arbitrary, but part of a pervasive pattern in which the Herero destiny and the 00001 firing are inextricably linked.

But *Gravity's Rainbow* of course is *fiction* (isn't it?). It is hardly surprising that the correspondences fit, since they were created to do so. What does this created pattern tell us about reality, if anything? The narrator recognizes that what he calls his "Kute Korrespondences" (p. 590) are the image, not the end point, of an infinite series hoping to "zero in on the tremendous and secret Function whose name [. . .] cannot be spoken" (p. 590). This is language in its fallen aspect, as an imperfect instrument that reveals only blurred and indistinct outlines of what was once perfect unity.

But the problem is insoluble only if we believe that unity is something that must be created, or more precisely, re-created. Consider the implications of the following progression: first language is seen as a process of recovery, an attempt to Return to the underlying deeper pattern. Then, language is an instrument not of recovery but of creation, actually bringing the patterns into existence. The next step is to recognize that the distinction between "reality" and "created pattern" is meaningless. This step implies a redefinition of the essential relationship between art and reality. Instead of asking, "What is breaking through the text as interface?" the appropriate question is, "What makes you think there is an interface?" In this view the elaborate metafictional machinery of the novel not only reflects but also challenges the perspective that we adopt when we operate within the subject-object duality.

In the Floundering Four episode, as the narrator's camera eye retreats from the stage on which the Floundering Four act to show a stadium full of spectators, among which are the Floundering Four, we are warned that "the Chances for any paradox here, really, are less than you think" (p. 680). The "monumental yellow structure" of the stadium is subject to a "never-sleeping percolation of life and enterprise through its shell, Outside and Inside interpiercing one another too fast, too finely labyrinthine, for either category to have much hegemony any more" (p. 681). When the subject-object duality is considered as an illusion that is imposed on reality rather than inherent in it, all we must do to recover the wholeness is to abandon the perspective that leads us to believe it is real. This provides one explanation for Slothrop's final dissolution. From a perspective that eschews the subject-object duality, for example from the perspective of Rilke's *Duino Elegies,* Slothrop has arrived at a transcendent realization of the essential connectedness of all things. In Zen terms, he has achieved satori, experiencing the self as a manifestation of the Universal One.[13]

But if to overcome the subject-object duality completely is to merge

[13]This is the argument Lance Ozier uses in "The Calculus of Transformation: More Mathematical Imagery in *Gravity's Rainbow,*" *Twentieth Century Literature,* 21 (1975), 193–210, to counter Joseph Slade's interpretation in *Thomas Pynchon* (New York: Warner Paperback Library, 1974) that Slothrop's dissipation implies he "never found himself" (p. 210). But to claim that Slothrop's dissipation is *simply* transcendence is to do violence to all of the negative connotations with which the narrator surrounds Slothrop's disappearance. Surely Mark Siegel is more nearly correct in pointing out that though in one sense Slothrop's dissipation is transcendence, in another sense "Slothrop has abandoned his ability to manipulate anything in the physical world" (*Pynchon: Creative Paranoia in Gravity's Rainbow* [Port Washington, N.Y., and London: Kennikat Press, 1978], p. 88.)

with the "mind-body" of the cosmos, it is also to cease to exist as a person localized in time and space, and thus to be unable to influence the temporal unfolding of events. Those who have made the transition, for example Walter Rathenau and Lyle Bland, cease to care how events unfold in secular history, for the good reason that to them secular history is an illusion. Something of the same indifference occurs in those who immerse themselves in the "mindless pleasures" that are the proletarian equivalent to Rilkean transcendence. If from one perspective this is salvation, from another point of view it is a betrayal of the revolution by becoming incapable of effective social action. To stay at the barricades, however, is to remain in the realm of cognitive thought, thus contributing, through the very act of remaining conscious, to Their enterprise.

I take this to be one meaning of Slothrop's "primal dream," in which he opens a German dictionary to find, opposite "JAMF," the definition "I" (p. 287). To try to fight Them is to become Them. The narrator has the "world-renowned analyst Mickey Wuxtry-Wuxtry" suggest that "there never was a Dr. Jamf [. . .] Jamf was only a fiction, [. . .] to help [Slothrop] deny what he could not possibly admit: that he might possibly be in love, in sexual love, with his, and his race's, death" (p. 738). The assumption that there could exist a "They-structure" distinct from a "We-structure" is exposed as an illusion at the novel's end when we split into the spectators watching a movie in which Gottfried falls in the Rocket, and the victims on whom the Rocket will land as it falls the "last delta-t." As long as we remain cognitively conscious, the holocaust is inevitable, and the realization that "everything is connected" leads only to the understanding that They are We.

Thus slowly, inexorably, Pynchon's text keeps returning to the central dilemma of how to speak from within a field without betraying it to the linear processes of articulation and cognition. The two conflicting impulses—the hope that Return to a pure apprehension of the field may be possible, and the recognition that such a hope is inherently contradictory—define the matrix within which the action of *Gravity's Rainbow* takes place.

The trajectory from which there is no Return can be called centrifugal; its metaphors are the Diaspora, the scattering of the seed, the Flight from the Center. "What if we're all Jews," Gwenhidwy tells Pointsman, "[. . .] all scattered like seeds? still flying outward from the primal fist so long ago" (p. 170). Pointsman, though he pretends to misunderstand, "knows what he means"; "he means alone and separate forever" (p. 170). Opposing this movement are the centripetal forces that extend beyond the boundaries of the individual self to identify with the life cycle as a whole. Säure Bummer prefers Rossini over Beethoven because "with Rossini, the whole point is that lovers always get together, isolation is overcome, and like it or not that is the one great centripetal movement of the World. Through the machineries of greed, pettiness, and the abuse of power, *love occurs*" (p. 440). The belief that "the World is rushing together" (p. 440) is the hope that alienation can be overcome by the simple forces of love and trust; it is identified with the possibility that there are routes back, paths of Return to the Center.

One scientific model Pynchon draws on to validate these two opposing impulses is the entropic decay implied by the Second Law of Thermodynamics, which opposes the tendency of life to create structure. Another scientific model that is equally important to Pynchon's scheme, and that more directly connects with the field concept, is the expanding model of the universe.[14] The study of how the cosmos began was revolutionized, as was so much else in physics, by Einstein's Special and General Theories of Relativity. It was from these two theories that modern cosmology was born. During the 1930s and 1940s, the implications of the field concept for cosmology were increasingly developed, and climaxed in the discovery of the cosmic background radiation that provided compelling evidence for the "Big Bang" theory of creation. According to the "Big Bang" model, at the beginning of time an unimaginably dense center containing everything in the universe exploded, expelling matter in every direction and creating, as it expanded, the present universe. In this model, the cosmos is conceived as coming into being at the expanding circumference of the initial explosion. Beyond the circumference lies an unimaginable void that lacks even the basic structure of spacetime, while at its center lies the

[14]The work of Ilya Prigogine on thermodynamic systems far from equilibrium has been seminal in elucidating the tendency of life to create structure; see for example Ilya Prigogine and G. Nicolis, *Self-Organization in Non-Equilibrium Systems: From Dissipative Structures to Order through Fluctuations* (New York: John Wiley, 1977). For a discussion of how seemingly low-entropy events can be reconciled with the Second Law of thermodynamics by considering the larger cosmological system within which they are contained, see P. C. W. Davies's discussion of "branch systems" in *The Physics of Time Asymmetry* (Berkeley and Los Angeles: University of California Press, 1977), pp. 68–74.

memory of what the narrator calls the "primal fist," the singularity that he recognizes as the "infinitely dense point from which the present Universe expanded" (p. 396). As Pynchon's narrator correctly asserts, this point of singularity is technically called a Friedmann point, after the "Russian mathematician [Alexander] Friedmann" (p. 396). These references suggest that Pynchon's mythic Flight from the Center is an imaginative reconstruction of the scientific model of an expanding universe. Like the fictional universe with which the characters in the narrative attempt to come to terms, it is invested by Pynchon with both a positive and negative valence, with the possibility that closure may be achieved and the possibility that it may not.[15]

The model's connotations of ambivalence come into focus through the text's treatment of "singularities." As we have seen, one example of a singularity is the Friedmann point, the hypothetical center from which the universe exploded. More generally, singularities are points in mathematical functions where the derivative, or rate of change, of the function becomes discontinuous.[16] One example of a singularity is a point where a function peaks sharply (Figure 1). In Pynchon's view, singularities pose a particular threat to the differential calculus because at a singularity the rate of change that the differential attempts to express goes to infinity. Figure 2 shows how, as the Δx increment approaches zero at a singularity, the Δy increment suddenly becomes very large. The differential, dy/dx, is defined as the limit, as Δx ap-

[15]For a good general description of how the "Big Bang" theory was developed and confirmed, see Timothy Ferris, *The Red Limit: The Search for the Edge of the Universe* (New York: William Murrow, 1977). Alexander A. Friedmann first proposed the existence of infinitely dense points (the "Friedmann point") in his paper, "Über die Drummung des Raumes," *Zeitschrift für Physik,* 10 (1922), 377–386. In this article Friedmann points out that an expanding universe that originated from a singularity in the spacetime matrix was in fact mathematically possible, in contradistinction to the steady-state model that Einstein has presupposed. Friedmann is commonly regarded as the father of the "Big Bang" theory of creation because of this work. (See also Friedmann's article "Über die Möglichkeit einer Welt mit konstanter negativer Krummung des Raumes," *Zeitschrift für Physik,* 21 [1924], 326–332.) The narrator's allusions to Friedmann reveal Pynchon's familiarity with these cosmological models.

[16]Lance W. Ozier in "The Calculus of Transformation" has some discussion of the singularity on pp. 202–204. However, in a diagram on p. 209 he seems inexplicably to associate a singularity with the mathematical operation for dividing a well-behaved function (in his diagram, a straight line) into smaller delta t increments. Actually, a singularity is what *disrupts* this differentiating process. Moreover, it is clear from Pynchon's repeated association of singularities with steeples, mountain peaks, etc. that he is using singularity in the sense of a sharp peak in the function.

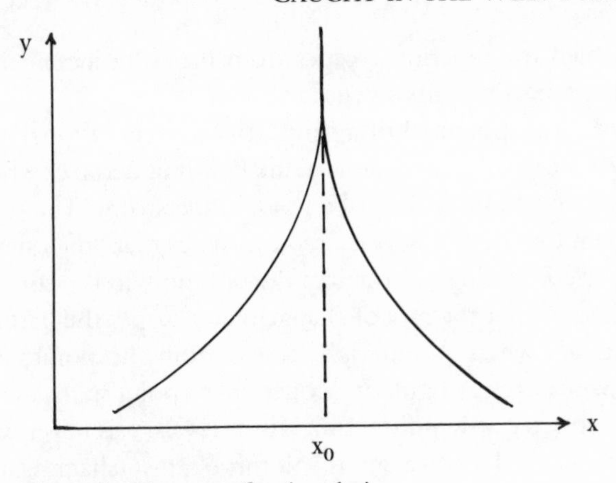

Figure 1. Diagram of a singularity

proaches zero, of $\Delta y/\Delta x$.[17] At the singularity, this limit must be formally expressed as infinity because it fails to converge, becoming larger and larger as the cusp is approached. The singularity thus represents a point where the behavior of the function ceases to be mathematically expressible, except in a purely formal way. Metaphorically, it is the

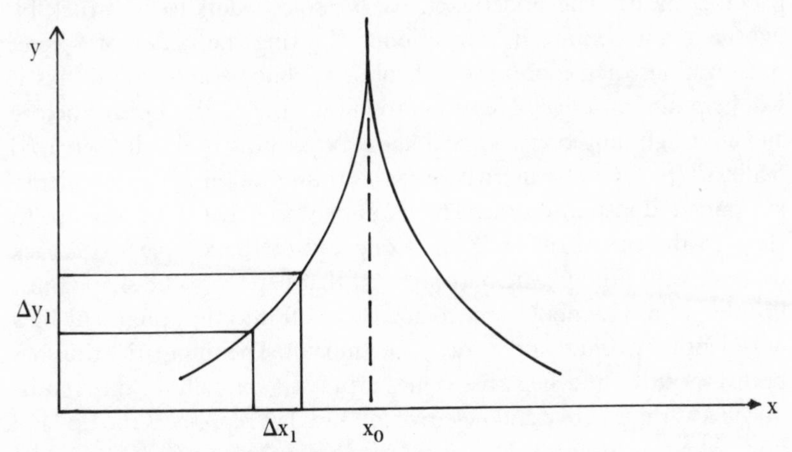

Figure 2. Differentiating a singularity

[17]Joseph Slade in *Thomas Pynchon* has a good diagram of this process of differentiation (p. 219).

point at which the function escapes from the delta increments of rational analysis into the unknown.

The mysterious potential of a singularity to defy rational analysis is the basis for the narrator's account of the Polish undertaker who, clad in a metal suit, rows out on the lake in a thunderstorm. The undertaker wants to be hit by lightning because he assumes that "the ones who do get hit experience a singular point, a discontinuity in the curve of life." "Do you know what the rate of change *is* at a cusp?" the narrator asks. "*Infinity,* that's what! A-and right across from the point, it's *minus* infinity" (p. 664). The singularity, concealing a point so mysterious that calculus, no matter how infinitesimal its intervals, can never capture it, is the mathematical equivalent to Slothrop's insouciant wanderings. When the narrator identifies the singularity with the steeple of the Empire State Building that King Kong climbs, he suggests that, like King Kong, Slothrop, or the Rocket, singularities possess the charismatic power to disrupt business-as-usual with their uncontrollable behavior.

But like all charismatic objects in *Gravity's Rainbow,* the singularity is subject to co-option. If it has the power to disrupt conventional modes of cognition, it can also become a tool in Their service, or twisted into paranoia by us. The undertaker, we are told, wants to be struck by lightning not because he cares about "busting the code" of "secret organizations or recognizable subcultures," but because "he thinks it will help him *in his job.* He wants to know how people behave before and after lightning bolts, so he'll know better how to handle bereaved families" (p. 665). The alternative to this routinization of the singularity is a paranoid response to it. The narrator warns that if we attempt to cling to the singularity as King Kong did to the steeple, "bareback dwarves with little plastic masks [. . .] that happen to be shaped just like the infinity symbol" wait to snatch us off into the congruent-but-not-identical world that the paranoiacs inhabit. The singularity thus has both a positive and negative value, expressed formally in differential calculus as the positive and negative infinity that represents the up- and down-slope at the cusp. The singularity has, in other words, a double-edged point.

What then are we to make of the narrator's emphasis on the Friedmann point, the singularity from which the universe began? Its double valence comes most clearly into view when we consider it as the point not only from which we began, but to which we will return. Though astronomers agree that the universe is expanding, they do not agree on whether this outward trajectory will continue forever. It is possible that the attractive forces between the masses that comprise the universe will eventually be able to overcome the outward movement. In this case the universe could begin contracting. The rush inward toward the Center would then end in another incredibly dense mass which would again explode, expelling matter outward. The universe would thus act like a rubber band being stretched and then released.[18]

In this case, the universe will not end in the heat death predicted by the Second Law of thermodynamics (and Pynchon's "Entropy"), but will continue to exist in unending cycles of Flight and Return that some see as a cosmic analogue to reincarnation. The physicist Thomas Gold has even suggested that in a contracting universe, entropy would spontaneously decrease.[19] The concerns that mark Pynchon's early fiction—the heat death of the universe, the Second Law of thermodynamics, the erosion of meaning that entropy implies—could thus be subject to qualification or reversal if the Universe can Return.

But the other edge of this point emerges with the realization that Return also means annihilation, for the universe can be reborn only by going through the absolute gravitational collapse that means not only the death of all life, but the destruction of all matter as we know it. This is the "Secret of the Fearful Assembly" (p. 738) that lies behind the narrator's various scenarios of Return, for example when he imagines the assembly of the 00001 as a "Diaspora running backwards, seeds of exile flying inward in a modest preview of gravitational collapse" (p. 737). The point of the 00001 Rocket, also called a "singularity," is another version of the Friedmann point, and thus implies both rebirth and annihilation. At the end of the text we get the point in both senses,

[18]For a discussion of the oscillating universe models see Davies, pp. 188–198. In an oscillating universe there would, strictly speaking, be no "Big Bang" or creation event, only the points at which one phase of the cycle ends and another begins.

[19]Thomas Gold, "The Arrow of Time," *American Journal of Physics,* 30 (1962), 403–410. See also Hermann Bondi, "Physics and Cosmology," *The Observatory,* 82 (1962), 133–143. It should be noted that Gold's arguments for the decreasing entropy of a contracting universe are open to serious objections; see, for example, Davies, pp. 96, 193ff. In order for entropy to decrease, the universe would somehow have to *anticipate* the pattern of Return so that the contraction would be ideally reversible in the thermodynamic sense (Davies, p. 199). That the odds against such an occurrence are staggeringly great is perhaps one reason why, in *Gravity's Rainbow,* the hope for Return is so qualified.

as that part of the Rocket about to penetrate our skulls and as the emergent meaning that links our comprehension of the larger, cosmic patterns of Return with the concomitant realization that they necessarily entail our personal annihilation. Thus the dilemma that has characterized all of Pynchon's representations of the field concept is writ large in the cosmological model: to remain conscious is to resist Return, and to Return is to experience the annihilation of consciousness that Slothrop foreshadows for us when he dissipates into the underlying field of the cosmos.

To this complex dynamic I should like to add one more complexity, in the form of a particularly bizarre singularity—the black hole. The crucial factor in whether the universe can stop expanding and begin contracting is the amount of mass it contains. Researchers have attempted to calculate this figure, but the numbers are so near to the critical mass that it is still too close to call; the answer could go either way. Those who believe there is enough mass to initiate return argue that some of it can't be seen because it is hidden in black holes. Many scoff at this conjecture, but it is difficult to disprove because no one understands exactly what happens in black holes; within their infinitely dense confines, the known laws of physics cease to be valid. The controversy is echoed in *Gravity's Rainbow,* and connects Pynchon's treatment of singularities with the larger questions posed by the narrator's attempt to create a narrative "field" in his text.

Black holes are thought to be created by the collapse of dying stars when the forces of gravitation become so intense that nothing can escape. Once matter or energy enters the circumference of this influence, called the "event horizon," all knowledge of that event is lost because nothing, not even light, will ever return to deliver information about it. The event horizon has a magnitude calculated by the Schwarzchild radius, named after Karl Schwarzchild, who noticed, in 1917, anomalies in Einstein's gravitational equations that later were recognized to describe black holes. Once a star contracts beyond its Schwarzchild radius, P. C. W. Davies writes, "the whole mass implodes to a zero volume and infinite density" (Davies, p. 98). In Pynchon's text, "Schwarzchild" is the Jamf code name for the infant Slothrop. The correspondence suggests that black holes are the charismatic objects in the scientific model that play a role analogous to Slothrop in the plot. In fact, black holes are recognized as singularities in the spacetime fabric

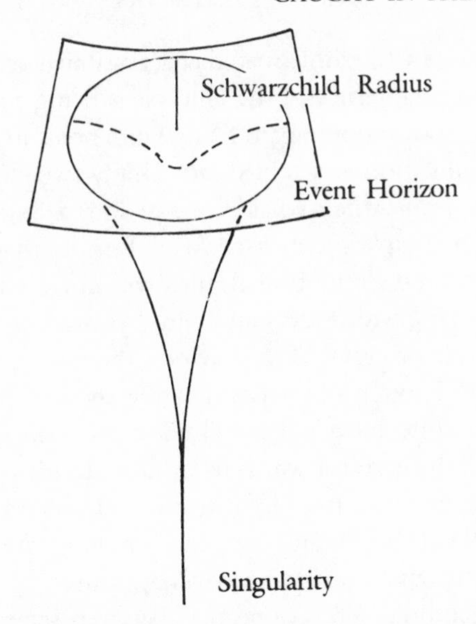

Figure 3. Deformation of spacetime by a black hole

of the universe; Figure 3 shows why their calculated shape justifies the name. Like other singularities, black holes too have a double-edged point, both a positive and negative value.

The doubleness is inherent in the equations predicting black holes, for it turns out that these equations have not one but two solutions. In the second solution, the equations yield a "white hole," a center from which energy and matter radiate outward rather than being sucked in as they are in a black hole. *Gravity's Rainbow* is filled with black-and-white images that are mirror reflections of one another and that can reverse into one another as they move through time; a black hole can be transformed into a white hole by reversing the value of time in the equations. The symbolic values of the two mirror images are also opposite. Whereas the black hole is a powerful metaphor for the absolute annihilation of no Return, the white hole promises rebirth through another "Big Bang." In von Göll's last cinematic production (shot, no doubt, in black and white), he runs the film backward to create a "reverse world" in which "the Great Irreversible is actually reversed as the corpse comes to life to the accompaniment of a backwards gunshot" (p. 745). Von Göll is not as deranged as he might seem, for the conversion of a black hole into a white hole would enact a similar scenario for the entire cosmos. If the substance of the universe is being sucked into black holes, it is being spewed out again from white holes in a circular dialectic in which annihilation and rebirth are simply two sides of the same coin. Taken as a *gestalt,* the two sides merge into a single picture of the cosmos itself participating in a cycle of Return that at once transcends and validates the attempts at Return within Pynchon's text.

We are now in a position to understand why Pynchon chose in his title to highlight the role of gravity, for gravity is the force connecting these two possibilities. Though physicists disagree about whether Return is possible, they concur on what will allow the contraction to begin. The gravity that the narrator warns us is "taken so for granted" is the elusive power that can turn the Flight from the Center around. If Return is possible, it will be because gravity is pulling the universe together again. But this same gravity is also what insures that nothing can return from a black hole. The connection between them helps to explain why gravity should be treated both as a Vice, an "old buffoon" leading us down the primrose path, and as a redemptive force controlling our destiny. Thus the enigma of gravity's double role is clarified when we realize it is the underlying force responsible both for the ultimate Flight from the Center that black holes represent, and the cosmic Return to that Center.

Not all of the enigmas, however, are susceptible to resolution. When the field models of cosmology define boundaries that are as wide, or as narrow, as an oscillating universe, the ambiguities that characterize human life are reinforced, not banished. If the universe can Return, then it provides a cosmic equivalent to the process of life, death, and new life that we see in Pirate's rooftop garden. Only by identifying with this universal cycle of Return can we find such comfort as the cosmic drama allows. The irony is that as long as we remain human, complete identification with it is impossible, for to be conscious is to value consciousness, and hence to resist annihilation. The narrator, after describing Lyle Bland's transcendent realization that Gravity is something "eerie" and "Messianic," mourns that

> the rest of us, not chosen for enlightenment, left on the outside of Earth, at the mercy of a Gravity we have only begun to learn how to detect and measure, must go on blundering in our front-brain faith in Kute Korres-

> pondences [. . .] kicking endlessly among the plastic trivia, finding in each Deeper Significance and trying to string them all together like terms of a power series hoping to zero in on the tremendous and secret Function whose name, like the permutated names of God, cannot be spoken . . . (p. 590)

Though the "Kute Korrespondences" between language and a field view of reality can provide a context for the questions *Gravity's Rainbow* poses, they cannot supply the answers. Like the real nature of gravity, our relation to the field of the cosmos is a "tremendous and secret Function" whose meaning finally "cannot be spoken," even by Pynchon.

REFERENCES CITED

IN LITERATURE

Alazraki, Jaime. "Tlön y Asterión: Metáforas Epistemológicas." In *Jorge Luis Borges*. Ed. Jaime Alazraki. No. 88 in *El Escritor y la Crítica*. Madrid: Taurus Ediciones, S.A., 1976.

Barrenechea, Ana Mariá. *La expresión de la irrealidad en la obra de Jorge Luis Borges*. Buenos Aires: Ed. Paidos, 1967. Translated as *Borges the Labyrinth Maker* by Robert Lima. New York: New York University Press, 1965.

Bersani, Leo. *A Future for Astyanax*. Boston and Toronto: Little, Brown, 1976.

Borges, Jorge Luis. *The Aleph and Other Stories, 1933–1969*. Ed. and trans. Norman Thomas di Giovanni in collaboration with Borges. New York: E. P. Dutton, 1970.

——. *Discusión. Obras Completas* vol. 6. Buenos Aires: Emecé Editores, 1957.

——. *Ficciones*. Ed. Anthony Kerrigan. New York: Grove, 1962.

——. *Historia de la Eternidad. Obras Completas,* vol. 1. Buenos Aires: Emecé Editores, 1953.

——. *El idioma de los argentinos*. Buenos Aires: n.p., 1928.

——. "Interview with Ronald Christ." *The Paris Review,* 40 (1967), 116–164.

——. *Other Inquisitions, 1937–1952*. Trans. Ruth L. C. Sims. Austin and London: University of Texas Press, 1964.

Butler, Colin. "Borges and Time: With Particular Reference to 'A New Refutation of Time.'" *Orbis Litterarum,* 28 (1973), 148–61.

Carlos, Alberto J. "Dante y 'El Aleph' de Borges." *Duquesne Hispanic Review,* 5 (1966), 35–50.

Christ, Ronald. *The Narrow Act: Borges' Art of Illusion*. New York: New York University Press, 1969.

Clarke, Colin. *River of Dissolution*. New York: Barnes and Noble, 1969.

Cowan, James C. *D. H. Lawrence's American Journey: A Study in Literature and Myth*. Cleveland: Case Western Reserve University Press, 1970.

Cro, Stelio. "Borges e Dante." *Lettere Italiane,* 20 (1968), 403–410.

Davison, Ned J. "Aesthetic Persuasion in 'A New Refutation of Time.'" *Latin American Literary Review,* 14 (1979), 1–4.

Dembo, L. S., ed. *Nabokov: The Man and His Work.* Madison: University of Wisconsin Press, 1967.

Eichner, Hans. "The Rise of Modern Science and the Genesis of Romanticism." *PMLA,* 97 (1982), 8–30.

Engelberg, Edward. "Escape from the Circles of Experience: D. H. Lawrence's *The Rainbow* as a Modern *Bildungsroman.*" *PMLA,* 78 (1960), 103–113.

Flower, Timothy F. "The Scientific Art of Nabokov's *Pale Fire.*" *Criticism,* 17 (1975), 223–233.

Friedman, Alan J., and Manfred Puetz. "Science as Metaphor: Thomas Pynchon and *Gravity's Rainbow.*" *Contemporary Literature,* 15 (1974), 345–359.

Gombrich, E. H. *Art and Illusion: A Study in the Psychology of Pictorial Representation.* Bollingen Series, vol. 25. Kingsport, Tenn.: Pantheon Books, 1960.

Harper, Howard, Jr. "*Fantasia* and the Psychodynamics of *Women in Love.*" In *The Classic British Novel.* Ed. Howard Harper, Jr., and Charles Edge. Athens, Ga.: University of Georgia Press, 1972.

Hinz, Evelyn. "The Beginning and the End: D. H. Lawrence's *Psychoanalysis and Fantasia.*" *Dalhousie Review,* 52 (1972), 251–265.

Hough, Graham. *The Dark Sun: A Study of D. H. Lawrence.* New York: Capricorn, 1959.

Irby, James E. *The Structure of the Stories of Jorge Luis Borges.* Ann Arbor: University of Michigan Microfilms, 1962.

Johnson, D. Barton. "The Ambidextrous Universe of Nabokov's *Look at the Harlequins!*" In *Critical Essays on Vladimir Nabokov.* Ed. Phyllis Roth. Boston: G. K. Hall, 1984.

——. "The Labyrinth of Incest in Nabokov's *Ada.*" *Contemporary Literature,* forthcoming.

Kermode, Frank. *D. H. Lawrence.* New York: Viking Press, 1973.

Lawrence, D. H. "The Crown." In *Phoenix II: Uncollected, Unpublished and Other Prose Works by D. H. Lawrence.* Ed. Warren Roberts and Harry T. Moore. New York: Penguin, 1978.

——. "Prologue to *Women in Love* (Unpublished)." Ed. George H. Ford. *Texas Quarterly,* 6 (1973), 98–111.

——. *Psychoanalysis and the Unconscious and Fantasia of the Unconscious.* Ed. Philip Reiff. New York: Viking Press, 1960.

——. *The Rainbow.* New York: Penguin, 1976.

——. *Women in Love.* New York: Penguin, 1976.

Leavis, F. R. *D. H. Lawrence, Novelist.* London: Chatto and Windus, 1955.

Levine, George, and David Leverenz. *Mindful Pleasures: Essays on Thomas Pynchon.* Boston: Little, Brown, 1976.

Levy, Salomon. "El *Aleph,* símbolo cabalístico, y sus implicaciones en la obra de Jorge Luis Borges." *Hispanic Review,* 44 (1976), 143–161.

Mason, Bobbie Ann. *Nabokov's Garden: A Guide to Ada*. Ann Arbor: Ardis, 1966.
Mendelson, Edward, ed. *Pynchon: A Collection of Critical Essays*. Englewood Cliffs, N.J.: Prentice-Hall, 1978.
Monegal, Emir, and Alistair Reid, eds. *Borges: A Reader*. New York: E. P. Dutton, 1981.
Nabokov, Vladimir. *Ada or Ardor: A Family Chronicle*. New York: McGraw-Hill, 1969.
——. "Anniversary Notes." Supplement to *TriQuarterly*, 17 (1970), 4–9.
——. *The Gift*. New York: Popular Library, 1963.
——. *Speak, Memory: A Memoir*. New York: G. P. Putnam's Sons, 1966.
"On the Road with Aristotle." *Times Literary Supplement*, April 19, 1974, pp. 405–406.
Ozier, Lance W. "The Calculus of Transformation: More Mathematical Imagery in *Gravity's Rainbow*." *Twentieth Century Literature*, 21 (1975), 193–210.
Pearce, Richard. "Enter the Frame." *TriQuarterly*, 30 (1974), 71–82.
Pirsig, Robert M. *Zen and the Art of Motorcycle Maintenance: An Inquiry into Values*. New York: Bantam, 1980.
Placher, William C. "The Trinity and the Motorcycle." *Theology Today*, 34 (1977), 248–256.
Purdy, Strother. *The Hole in the Fabric*. Pittsburgh: University of Pittsburgh Press, 1977.
Pynchon, Thomas. *Gravity's Rainbow*. New York: Viking Press, 1973.
Rodino, Richard H. "Irony and Earnestness in Robert Pirsig's *Zen and the Art of Motorcycle Maintenance*." *Critique*, 22 (1980), 24.
Saussure, Ferdinand de. *Course in General Linguistics*. Ed. Charles Bally and Albert Sechehaye, trans. Wade Baskin. New York: Philosophical Library, 1959.
Siegel, Mark. "Creative Paranoia: Understanding the System of *Gravity's Rainbow*." *Critique*, 13 (1977), pp. 39–54.
——. *Pynchon: Creative Paranoia in Gravity's Rainbow*. Port Washington, N.Y., and London: Kennikat Press, 1978.
Slade, Joseph. *Thomas Pynchon*. New York: Warner Paperback Library, 1974.
Stark, John. "*Zen and the Art of Motorcycle Maintenance*." *Great Lakes Review: A Journal of Midwest Culture*, 3 (1977), 50–59.
Steele, Thomas S. "Zen and the Art . . . : The Identity of the Erlkönig," *Ariel*, 10 (1978), 83–93.
Steiner, George. "Uneasy Rider." *The New Yorker*, April 15, 1974, pp. 149–150.
Tanner, Tony. *City of Words: American Fiction, 1950–1970*. New York: Harper & Row, 1971.
Teilhard de Chardin, Pierre. *The Phenomenon of Man*. Trans. Bernard Wall. New York: Harper & Brothers, 1959.
Weber, Frances Wyers. "Borges' Stories: Fiction and Philosophy." *Hispanic Review*, 36 (1968), 124–141.

Westervelt, Linda. "'A Place Dependent on Ourselves': The Reader as System-Builder in Gravity's Rainbow." *Texas Studies in Literature and Language*, 22 (1980), 67–90.

Whorf, Benjamin. *Collected Papers on Metalinguistics*. Washington, D.C.: Foreign Service Institute, 1952.

Wittgenstein, Ludwig Joseph Johann. *Tractatus Logica-Philosophicus*. Ed. D. F. Pear and B. F. McGuinness. London: Routledge & Paul, 1961.

Wolfley, Lawrence C. "Repression's Rainbow: The Presence of Norman O. Brown in Pynchon's Big Novel." *PMLA*, 92 (1977), 873–889.

Zeller, Nancy Anne. "The Spiral of Time in *Ada*." in *A Book of Things about Nabokov*. Ed. Carl R. Proffer. Ann Arbor: Ardis, 1974.

IN SCIENCE

Bohm, David. "The Implicate Order: A New Order for Physics." *Process Studies*, 8 (1978), 73–101.

——. *Wholeness and the Implicate Order*. London and Boston: Routledge & Kegan Paul, 1980.

Bondi, Hermann. "Physics and Cosmology." *The Observatory*, 82 (1962), 133–143.

Cantor, Georg. "Beiträge zur Begründung der transfiniten Mengenlehre." Part 1. *Mathematische Annalen*, 46 (1895), 481–512.

——. "Beiträge zur Begründung der transfiniten Mengenlehre." Part 2. *Mathematische Annalen*, 49 (1897), 207–243.

Capek, Milic. *The Philosophical Impact of Contemporary Physics*. Princeton: D. Van Nostrand, 1961.

Capra, Fritjof. *The Tao of Physics: An Exploration of the Parallels between Modern Physics and Eastern Mysticism*. New York: Bantam Books, 1977.

Colodny, Robert G. *Problems and Paradoxes: The Philosophical Challenge of the Quantum Domain*. Pittsburgh: University of Pittsburgh Press, 1972.

Dauben, Joseph Warren. *Georg Cantor: His Mathematics and Philosophy of the Infinite*. Cambridge: Harvard University Press, 1979.

Davies, P. C. W. *The Physics of Time Asymmetry*. Berkeley: University of California Press, 1977.

Davis, Martin. "What Is a Computation?" In *Mathematics Today: Twelve Informal Essays*. Ed. Lynn Sheen. New York and Berlin: Springer-Verlag, 1978.

Einstein, Albert. *Autobiographical Notes*. Trans. and ed. Paul Arthur Schilpp. La Salle, Ill.: Court, 1979.

——. *Ideas and Opinions*. Ed. Carl Seelig. New York: Dell, 1973.

Ferris, Timothy. *The Red Limit: The Search for the Edge of the Universe*. New York: William Murrow, 1977.

Feynman, Richard P., Robert B. Leighton, and Matthew Sands, *The Feynman Lectures on Physics*, vol. 3. Reading, Mass.: Addison-Wesley, 1963–1965.

Freud, Sigmund. "Beyond the Pleasure Principle." *The Standard Edition of the*

Complete Psychological Works of Sigmund Freud, trans. James Strachey, vol. 18. London: Hogarth Press and the Institute of Psycho-Analysis, 1955.

Friedmann, Alexander A. "Über die Krümmung des Raumes." *Zeitschrift für Physik*, 10 (1922), 377–386.

——. "Über die Möglichkeit einer Welt mit konstanter negativer Krümmung des Raumes," *Zeitschrift für Physik*, 21 (1924), 326–332.

Gardner, Martin. *The Ambidextrous Universe*. New York: Charles Scribner's Sons, 1964.

——. *The Ambidextrous Universe: Mirror Asymmetry and Time-Reversed Worlds*. 2d ed. New York: Charles Scribner's Sons, 1979.

——. "Can Time Go Backward?" *Scientific American*, 216 (1967), 98–108.

——. *The Relativity Explosion*. Rev. ed. New York: Vintage Books, 1976.

Gold, Thomas. "The Arrow of Time." *American Journal of Physics*, 30 (1962), 403–410.

Hanson, N. R. *Patterns of Discovery*. Cambridge: Cambridge University Press, 1958.

Heijenoort, J. van. *From Frege to Gödel: A Source Book in Mathematical Logic*. Cambridge: Harvard University Press, 1967.

Heisenberg, Werner. *Physics and Beyond: Encounters and Conversations*. New York: Harper & Row, 1971.

——. *Physics and Philosophy: The Revolution in Modern Science*. New York: Harper & Row, 1958.

Hofstadter, Douglas R. *Gödel, Escher, Bach: An Eternal Golden Braid*. New York: Basic Books, 1979.

Jammer, Max. *The Philosophy of Quantum Mechanics*. New York: John Wiley, 1974.

Jourdain, P. E. B. *Contributions to the Founding of the Theory of Transfinite Numbers*. Chicago: Open Court, 1915.

Jung, C. G. *Synchronicity: An Acausal Connecting Principle*. Trans. R. F. C. Hull. Bollingen Series, vol. 20. Princeton: Princeton University Press, 1973.

Kasner, Edward, and James Newman. *Mathematics and the Imagination*. New York: Simon & Schuster, 1940.

Keller, Evelyn Fox. "Cognitive Repression in Contemporary Physics." *American Journal of Physics*, 47 (1979), 718–721.

——. "Gender and Science." *Psychoanalysis and Contemporary Thought*, 1 (1978), 409–433.

Kline, Morris. *Mathematics: The Loss of Certainty*. New York: Oxford University Press, 1980.

Kuhn, Thomas. *The Structure of Scientific Revolutions*. 2d ed. Chicago: University of Chicago Press, 1970.

Lakatos, I., and A. Musgrave, eds. *Criticism and the Growth of Knowledge*. Cambridge: Cambridge University Press, 1970.

Lanczos, Cornelius. *Albert Einstein and the Cosmic World Order*. New York: Interscience Publishers, 1965.

Lee, T. D. "Space Inversion, Time Reversal and Particle-Antiparticle Conjugation." *Physics Today,* 19 (1966), 23–31.

Lorentz, H. A., A. Einstein, H. Minkowski, and H. Weyl. *The Principle of Relativity.* New York: Dover, 1923.

Morin, Edgar. "Beyond Determinism: The Dialogue of Order and Disorder." *Sub-Stance,* 40 (1983), 22–35.

Neisser, Ulric. *Cognitive Psychology.* Englewood Cliffs, N.J.: Prentice-Hall, 1967.

Petersen, Aage. "The Philosophy of Niels Bohr." *Bulletin of the Atomic Scientists,* 19 (1963), 10–11.

Polanyi, Michael. *Personal Knowledge.* Chicago: University of Chicago Press, 1958.

——. *Science, Faith and Society.* Chicago: University of Chicago Press, 1964.

Prigogine, Ilya, and G. Nicolis. *Self-Organization in Non-Equilibrium Systems: From Dissipative Structures to Order through Fluctuations.* New York: John Wiley, 1977.

Russell, Bertrand. *The ABC of Relativity.* Rev. Ed. Fair Lawn, N.J.: Essential Books, 1958.

Snapper, Ernst. "The Three Crises in Mathematics: Logicism, Intuitionism and Formalism." *Mathematics Magazine,* 52 (1979), 207–216.

Suppe, Frederick, ed. *The Structure of Scientific Theories.* 2d ed. Urbana: University of Illinois Press, 1977.

t'Hooft, Gerard. "Gauge Theories of the Forces between Elementary Particles." *Scientific American,* 243 (1980), 104–137.

Toda, M. "Time and the Structure of Human Cognition." In *The Study of Time II.* Ed. J. T. Fraser and N. Lawrence. New York, Heidelberg, and Berlin: Springer-Verlag, 1975.

Weyl, Hermann. *Philosophy of Mathematics and Natural Science.* Princeton: Princeton University Press, 1949.

Whitehead, Alfred North, and Bertrand Russell. *Principia Mathematica.* 3 vols. 2d ed. New York: Cambridge University Press, 1925–27.

Wigner, Eugene P. "Violations of Symmetry in Physics." *Scientific American,* 213 (1965), 28–36.

INDEX

Library of Congress Cataloging in Publication Data

Hayles, N. Katherine.
The cosmic web.

Bibliography: p.
Includes index.
1. Literature, Modern—20th century—History and criticism. 2. Literature and science. I. Title.
PN771.H36 1984 809'.93356 84-45141
ISBN 0-8014-1742-2